NATURAL DISCOURSE

NATURAL DISCOURSE

Toward Ecocomposition

SIDNEY I. DOBRIN

CHRISTIAN R. WEISSER

STATE UNIVERSITY OF NEW YORK PRESS

This one's for
Jenna, Jaylen, Roy V, Remi, Daniel, and Eva

•

This book is for the chiefs of my tribe:
Robert V. and Robert C. Weisser

•

This book is dedicated to the world's oceans

All authors' royalties earned from sales of this book will be donated to
The Surfrider Foundation to help provide education about and
protection for the world's oceans and beaches. To learn more about
The Surfrider Foundation, visit http://www.surfrider.org or write to:

The Surfrider Foundation
122 S. El Camino Real PMB #67
San Clemente, CA 92672

Published by
State University of New York Press, Albany

© 2002 State University of New York

For information, address State University of New York Press,
90 State Street, Suite 700
Albany, NY 12207

Production by Michael Haggett
Marketing by Michael Campochiaro

Library of Congress Cataloging-in-Publication Data

Dobrin, Sidney I., 1967–
Natural discourse: toward ecocomposition /Sidney I. Dobrin, Christian R.
Weisser.
    p.   cm.
Includes bibliographical references and index.
ISBN 0–7914–5355–3 (alk. paper)— ISBN 0–7914–5356–1 (pbk.: alk. paper)
1. English language—Rhetoric—Study and teaching. 2. Natural history
literature—Authorship—Study and teaching. 3. Environmental literature—
Authorship—Study and teaching. 4. Natural history—Authorship—Study and
teaching. 5. Ecology—Authorship—Study and teaching. 6. Interdisciplinary
approach in education. 7. Academic writing—Study and teaching. 8. Nature—
Study and teaching. I. Weisser, Christian R., 1970. II. Title.
PE1479.N28 D63 2002
808'.066333—dc21   2001054943

10 9 8 7 6 5 4 3 2 1

Contents

Acknowledgments

Since writing is an ecological endeavor, and our writing exists in a tangle of relationships with other texts, authors, and conversations, we would like to thank and acknowledge all of those who helped us complete this work. We are especially grateful to Anis Bawarshi, Julie Drew, and Chris Keller for their ongoing suggestions and input. We are grateful to Carla Blount for her support and assistance. We are deeply indebted to M. Jimmie Killingsworth, John Krajicek, Derek Owens, and Randall Roorda for their advice, support, and continuing encouragement. We wish to acknowledge and thank Ed White for contributing his wonderful forward to this book and for his never-ending support. We would like to thank Gary Olson for his continual encouragement and his talent for doubting us enough to push us toward higher aspirations. Thanks to Michael Haggett and Michael Campochiaro of SUNY Press for their work on this project. We are, of course, forever in debt to Priscilla Ross for her support of this project. We also owe a special debt of thanks to photographer David Pu'u, who donated the cover photograph for this book. And, as a good deal of this book was conceived and written on board a number of fishing and dive boats, we wish to thank all of those boat captains who told us to quit talking about discourse and either get in the water or reel in a fish.

Foreword

EDWARD M. WHITE

Our oldest writings deal with the intersection of humans and their environments: Adam and Eve harvesting more than they ought in the Garden of Eden, Odysseus trying to get home despite the winds and the waves, and so on. Ecocomposition may be a new term but it speaks to the oldest of human concerns. However, as Sidney Dobrin and Christian Weisser point out, something new has been added to this oldest of issues: we have recently gained, and actually begun to use, the power to destroy our environment and most life on our planet along with it. Thus the sense of urgency that we find throughout this book. Is there anything that matters more?

This is not just one more composition textbook, nor even less, one more composition topic. Ecocomposition asks us above all to put ourselves in relation to the Mother of all topics, Nature herself. As Adam and Eve learned to their, and our, sorrow, to exploit Nature for personal goals is to lose out Big Time, to lose the intimate connection to Nature that allows us to live in comfort with Her. Or perhaps to live at all.

Every student, to be sure, has written about the environment, and deplored the sad state of affairs in far-off shrinking rain forests and expanding deserts. But that is not really what this book is about. Rather than evoke such routine complaints, Ecocomposition seeks to change the way we think about writing, and hence the way we think about the world we live in. For this is a book about rhetoric and deep theory of composition as much as it is about the natural environment. Here you will find analyses of environmental rhetoric, entirely new ways of considering the means of argument, original ways of reconsidering the fundamentals of composition. Logos, pathos, and ethos take on new meanings in the context of environmental discourse. The authors ask us not only to think in new and more profound ways about Nature but also to understand writing as "natural discourse," whose goals are to help writer and reader deepen their understanding of the interaction of humans with our world. This is an environmental book that is out to change the nature of writing instruction through a new view of the writing process.

Writing, as everyone knows, is a difficult craft to control. But we are mistaken if we think that understanding Nature or even thinking about Nature is an easy matter. How much effort it takes the modish to achieve the natural look! A character in Jane Austen's novel *Emma* (1816) remarks that the most natural way for English ladies and gentlemen to dine is indoors, with their servants. From the first page of *Natural Discourse,* the authors force us to confront the paradox that Nature is not simply "out there" but rather a construct of human invention, human behavior, human language. They point out that "no matter how lost in the wild one tries to get, the natural environment is a world constructed and defined by human discourse." Later on, they expand their claim, arguing that "the environment is an idea that is created through discourse." Since humans inhabit both a "biosphere" (our physical world) and a "semiosphere" (consisting of discourse, which shapes our existence and allows us to make sense of it), the project of this book is to bring both spaces together, to show that they are functions of each other.

Dobrin and Weisser are ambitious indeed. They intend no less than to shake the foundations of composition studies and its theoretical roots in classical rhetoric, to resituate the foundational course for university study in ecology. They see such competing notions as cultural studies or identity formation as coming under the umbrella of ecocomposition, since the notion of place precedes all others: "ecocomposition posits that environment precedes race, gender, and culture." Indeed, "ecocomposition should be seen not as a way one kind of (nature) writing gets done, but as how all writing is produced."

Finally, what surprises and intrigues me most about this book is its range. I had expected, from the title, an inquiry into a rather narrow topic. But the book keeps expanding its concerns, though always focused on its twin centers, ecology and composition. The authors maintain, quite properly, that they are, after all, compositionists, and their principal interest is to explore the most basic roots and most fundamental properties of this new and most exciting field of study. But the book ranges across the scholarship of the entire field, seeking to situate all aspects of composition as housed or to be housed in ecocompostion. If their project is successful, composition studies will never again be quite the same.

Flagstaff, Arizona: May 7, 2001

CHAPTER I

Ecocomposition

≫⋈≪

*Life must become more than
the wants and needs of humans.
We are not on this earth alone.
If the manatee goes into extinction
because the needs of humans
became more important
then we will have taken another
step backward toward our own
demise. Extinction is forever
and for all.*
　　　　　—Dr. Harvey Barnett

We begin this book with these words not for what they say—though their
message is important—but because of where one must go to read them in
the original: 22 feet below the surface of Crystal River in Florida, where
endangered manatees spend the winter. These words appear in a place
where words might seem a foreign thing, an intrusive thing in a natural
place.[1] Yet, when we dive at Crystal River, as we often have, the words re-
mind us of how enmeshed the world of words, of text, and the natural
world are. They are a reminder that human hands have mapped and defined "natural" places and that no matter how lost in the wild one tries to
get, the natural environment is a world constructed and defined by human
discourse. Even in the most remote portions of the Everglades—the quin-
tessentially unique natural Florida environment—does one find written
text warning of contaminated water, of human intrusion in wild places.
Even the very boundaries of the Everglades are mapped and legislated
through text; where nature may exist is discursively regulated. The very act

of naming a place "Everglades" distinguishes what is and what is not of that environment.

Such examples are by no means limited to Florida. Half a world away on the "Big Island" of Hawaii, words demarcate where natural sites exist. Each year, thousands of tourists flock to Hawaii Volcanoes National Park to get a glimpse at a "real" volcano. Most of them stop at the small town just outside the park, aptly named "Volcano," to purchase T-shirts, coffee mugs, and postcards emblazoned with images of Mauna Loa or Kilauea. Little attention is paid to the fact that it is *all volcano,* that the Hawaiian Islands themselves are volcanic peaks formed over the course of millions of years. In fact, this mapping and classifying of environments extends even below the depths of the world's oceans, encompassing sites as yet inaccessible to human beings. Fifteen miles south of Hawaii, more than three thousand feet below sea level, lies the volcanic seamount *Loihi.* Despite the fact that it will not reach sea level for tens of thousands of years, tour guides, residents, and even some scientists have begun referring to *Loihi* as the *next* Hawaiian island, thereby mapping and inscribing a long political, cultural, and social history upon this island that is yet to be.

Relationships between text and nature are impossible to avoid. In fact, postmodernity has come to identify nature as text, despite the fact that humans often ascribe anthropomorphic languages to that text rather than listening to or reading nature's own text. It is our goal to explore the relationships between discourse and natural systems, between language and environment, and between writing and ecology. For just as we spend much of our recreational time under or on the water or rambling in natural places, our professional time is spent examining discourse and teaching composition. And though, at first, the worlds of environmental and ecological thinking and composition scholarship and pedagogy might seem only remotely related, it is our agenda to show that not only are environment and composition closely bound to one another, but that the work of composition studies is an ecological endeavor. We wish to show that ecocomposition is a critical part of our scholarly inquiry in composition studies. We agree with Kenneth Burke (1965) who argues that intellectual life cannot be removed from "life," from biological, natural existence.

Environmental issues have become predominant in political and scholarly conversations in the late 1990s and early 2000s. American universities have begun granting degrees in Ecology, Wildlife Management, and Environmental Engineering as well as many other environmentally and ecologically based areas of study. However, until very recently, most academic endeavors regarding environmental and/or ecological concerns have been addressed primarily in the "hard sciences." In fact, composition, and much of the rest of the humanities has been resistant to scientific inquiry (we will discuss this resistence later). In the latter half of the 1990s, research, scholarship, and knowledge-making in academia began to redefine and extend disciplinary boundaries in order to provide more contextual, holistic, and useful ways of examining the world. Likewise, scholars have begun to inquire as to how environmental issues impact art, literature, discourse, and other areas of interest to scholars in the humanities. Yet, composition and rhetoric's inclusion of the "hard sciences" in its interdisciplinary agenda has been limited for the most part to cognitive psychology. There has been, as we have said, a resistence to the methodologies employed by the sciences. And while certainly composition studies and postmodern humanities studies need to question scientific method and inquiry—as do postmodern scientists Sandra Harding, Donna Haraway, and others—ecocomposition also identifies a need to turn to ecological methodologies in our study of written discourse. While neither of us claim to be scientists, or even experts in the ecological, environmental, or natural sciences, we do identify that what these scientific inquiries provide can be of great use to composition studies.

A new orientation toward ecological notions of composition—conceptually, methodologically, metaphorically, and pedagogically—promises to be an exciting and potentially meaningful direction for the field of composition studies. Like many of the other important movements in writing instruction over the past forty years, the move toward ecocomposition stands to develop more sophistication and complexity by incorporating research, theories, and scholarship in other academic disciplines. Just as cognitivist approaches to composition were influenced by inquiries into psychology, social constructionist perspectives were influenced by work in philosophy and the social sciences, and more recent post-process investigations have been influenced by cultural studies, social theory, feminism, and postcolonial studies, we feel that composition can also benefit from work in ecology and environmental sciences. In other words, it is only *natural* that composition studies recognizes its affiliation with ecological and environmental disciplines, and it stands to reason that our understandings of *discourse* can only become more precise and sophisticated through investigations that recognize the importance of these studies.

Composition studies is certainly not unique in its desire to turn to interdisciplinary investigations in order to extend and elaborate its understandings of a subject. Granted, most scholarship in the two fundamental academic domains, commonly called the "humanities" and the "sciences," still has a look of permanence to it. At times, the gaps between these two domains, and often even between disciplines that fall within the same domain, seem insurmountable. The various disciplines have their own languages, codes, theories, and jargonistic terms, and as a result, they are often unable or unwilling to communicate with one another. Ecocomposition hopes to bridge this gap between domains by recognizing that the specialization of composition studies—discourse—is inextricably linked to at least one specialization in the hard sciences—ecology. Perhaps one of the most significant goals of ecocomposition is its desire to cross the boundaries between the two academic cultures of the humanities and the sciences, and, in the process, make the connections between the various tongues of each. "This polarization [between the humanities and the sciences] is sheer loss to us all," wrote C. P. Snow in his defining 1959 essay *The Two Cultures and the Scientific Revolution*. "To us as a people, and to our society. It is at the same time practical and intellectual and creative loss." To use a spatial metaphor, ecocomposition considers the crossing of distances between the sciences and the humanities to be of utmost importance. As the well-known biologist Edward O. Wilson writes in *Consilience: The Unity of Knowledge,*

> There is only one way to unite the great branches of learning and end the culture wars. It is to view the boundary between the scientific and literary cultures not as a territorial line but as a broad and mostly unexplored terrain awaiting cooperative entry from both sides. The misunderstandings arise from ignorance of the terrain, not from a fundamental difference in mentality. (137)

Ecocomposition explores this terrain in an effort to formulate better understandings of discourse and its relationship to the world we live in. However, we'd like to avoid the conquestatory and acquisitive metaphors that Wilson seems to suggest, since these metaphors reflect a mindset that is at the root of the current ecological crisis. As we will show later in detail, the discursive construction of the natural world has been (at least since the Enlightenment) used to justify its exploitation. We prefer to view the role of ecocompositionists as intellectual travelers who explore new territories in an effort to change themselves, taking nothing but experiences and knowledge, preserving the integrity and resources of the space traveled in for others to experience and learn from. Ecocompositionists delve into ecological and environmental studies not to extend our territory in the intellectual landscape, but to improve our understanding of the connections between these related disciplines, discourses, and epistemologies.

In the last twenty-five years, theorists and researchers in the social sciences and humanities have embraced the systematic exploration of social relations and culture as integral to the study of the construction of knowledge (epistemology). Likewise in composition studies, the social dimensions of language have dominated scholarly conversations concerning the construction of knowledge. That is, in the 1980s, many composition theorists and researchers began to focus on the social nature of writing and suggested that the correlation between social experience and writing ability is palpable. This orientation had widespread implications for composition theory, and brought with it, for example, new ways of thinking about an individual's identity (very often, the student in a writing class) and how identity is manifested through writing and speaking. As Christian has suggested in his essay "Ecocomposition and the Greening of Identity," social constructionist approaches to composition "expanded the way we thought of identity, asserting that it emerges not just from the internal processes of the individual, but also from a wider variety of influences: the social conventions we share with other human beings" (83–84). Within the past decade, compositionists have focused much of their attention post-process toward the critical categories of race, gender, class, and culture. These beneficial inquiries have aided in continuously redefining ways in which language impacts human thought and identity. Within just the past few years, some compositionists have begun to include place and environment as other critical categories in this very inquiry. All of the essays in the collection we recently edited, *Ecocomposition: Theoretical and Pedagogical Perspectives* (2000), recognize the importance of ecological approaches to composition. This recognition is long overdue, and the inclusion of ecological and/or environmental perspectives in composition theory is essential to the discipline's continued growth and development. As Cheryll Glotfelty explains, "If your knowledge of the outside world were limited to what you could infer from the major publications of the literary profession, you would quickly discern that race, class, and gender were the hot topics of the late twentieth century, but you would never suspect that the earth's life support systems were under stress. Indeed, you might never know that there was an earth at all" (xvi). Though leveled as a critique of literary criticism, the same critique applies to composition studies. A perusal of the major journals and publications in composition studies of the last decade reveal composition's turn toward issues of cultural studies, post-process writing, and various other socially based issues of discourse. Yet, with very limited exceptions in the last few years of the 1990s, compositionists have been wary of addressing issues of ecology, environment, place, location, and habitat in their scholarship. More recently, however, composition-specific publications such as *JAC: A Journal of Composition Theory, Composition Studies,* and *College Composition and Communication* have begun to print a very limited number of environment-directed articles that specifically address the intersections between composition and environmental concerns. We hope to promote and advance the importance of examining the intersections between discourse, place, and environment through theoretical examinations and pedagogical approaches and to explain how and why composition's roots do indeed tap into ecological sciences in their current incarnations. We hope to show how the two massive cultural projects of composition studies and ecology might inform one another and to identify how composition studies is very much an ecological inquiry. That is, we offer here a call to composition to embrace the work being done in ecocomposition.

At this point, it is important to establish a working definition of exactly what (we feel) ecocomposition comprises. We provide this definition as a conceptual framework, a ground in which more fruitful, complex studies might emerge. We do not intend to offer here in this rudimentary definition, or anywhere else in this book, a final definition that excludes other interpretations. We hope only to provide a point of origin for others who wish to travel similar terrain. We would also like to establish that in this book, when we say things like "ecocomposition is" or "ecocomposition contends," we do so working from our current conceptions of the subject, which result from a still rudimentary phase in ecocomposition's development. We are certain that ecocomposition will flourish in productive and worthwhile directions, and we hope only that this study will facilitate this evolution. That being said, we offer the following definition of ecocomposition:

> Ecocomposition is the study of the relationships between environments (and by that we mean natural, constructed, and even imagined places) and discourse (speaking, writing, and thinking). Ecocomposition draws primarily from disciplines that study discourse (chiefly composition, but also including literary studies, communication, cultural studies, linguistics, and philosophy) and merges the perspectives of them with work in disciplines that examine environment (these include ecology, environmental studies, sociobiology, and other "hard" sciences). As a result, ecocomposition attempts to provide a more holistic, encompassing framework for studies of the relationship between discourse and environment.

In the pages that follow, we offer further working definitions of various terms and theories pertinent to the evolution of *ecocomposition*. We explore the history of environmental, ecological, conservationist, and preservationist writing and their recent emergences in the humanities and the sciences, tracing the history of the discussions surrounding them in order to establish a theoretical rationale for ecocomposition. We explore the liberatory, activist potentials of environmental and ecological discourse, and we develop pedagogical approaches for the ecocomposition class-room. Within this inquiry, we also hope to provide perspectives as to why environmental studies are crucial to our work as compositionists and our lives as human beings and to highlight ways in which contemporary composition studies is already ecological.

Beginning Places

We begin here by establishing the context of *ecocomposition*. We have heard the word used loosely in several instances over the past few years, though we have seen no attempt at defining the agendas of this scholarly inquiry. In their 1998 panel at the Conference on College Composition and Communication Convention in Chicago, Randall Roorda, Lee Smith, and Michael McDowell began to introduce *ecocomposition* to the composition population. None offered any formal definition for the term *ecocomposition* as such. Roorda's introductory remarks can be seen as landmark in the evolution of ecocomposition as he called for the Association for the Study of Literature and Environment to redirect their focus from "Literature" to "Literacy." In this call, Roorda offered a definition of "ecological literacy" based on David W. Orr's definition of the same term. Roorda linked ecological literacy with the process orientation in composition to introduce a "process-oriented aspect of literacy and environmental education" to composition studies. Roorda then moved to establish the ASLE-CCCC Special Interest Group which met for the first time the following year at CCCC in Atlanta. We are grateful for Roorda's effort to create a site for ecocomposition in CCCC, yet at no time did Roorda's introductory remarks and actions move toward defining ecocomposition as a school of thought, inquiry, or critical approach which could be situated in composition studies.

For Lee Smith, the assumption was that ecocomposition was a term of familiarity and her focus turned to how ecocomposition and service learning might interact in a classroom. Her talk highlighted an interesting course design which linked "ecocomposition and service learning" in order to "provide opportunities for students to arrive at [environmental] awareness and engage in real world research and writing." Certainly, service learning and real world writing assignments are a part of ecocomposition, and Smith's talk helped promote thinking about how ecocomposition pedagogies might evolve, but it lacked the definitions we sought for ecocomposition. Similarly, Michael McDowell provided a wonderful discussion of how the spaces of computer-assisted classrooms can be examined as ecological spaces. McDowell argued that "Ecocomp terminology and ideas not only help explain what happens as our students write, but ecocomp also offers a necessary antidote to some of the usually ignored negative effects of using computers in composition classes." Yet, like Smith, McDowell leaves "ecocomp" undefined and assumed in his CCCC talk.

In his essay "Talking about Trees in Stumptown: Pedagogical Problems in Teaching EcoComp," McDowell also provides a substantial and important discussion of developing ecocomposition pedagogies. He correctly claims that "environmental issues make ideal subjects for composition classes because they are as complex, as multidisciplinary, and as emotionally charged as any social issues can be: they are based upon cultural assumptions that are currently changing; every student has direct personal experience with them; and many environmental issues engage every sense we have" (19). McDowell then describes the types of problems that may arise in an ecocomposition course and goes on to provide helpful solutions to some of these problems. McDowell also assesses the various textbooks designed for an ecocomposition classroom, and then discusses ways in which local issues may be brought into the ecocomposition class. It is only then that McDowell provides any definition of ecocomposition.

> Although now I only recommend a daily newspaper, the inclusion of daily environmental news into the class has expanded the definition of EcoComp to embrace almost every subject a student can think of. Whereas initially I thought of these courses as focusing on the natural world, now they focus on the environmental aspect of any significant issue, the closer to home the better. (22)

While we certainly agree with McDowell's initiative to involve local texts and to encourage the exploration of environmental aspects of any event, we also acknowledge the limited and limiting view of ecocomposition which this definition provides. First, and foremost, we want to identify that ecocomposition—both an inquiry and a pedagogy—does not simply focus on the natural world. As we suggested in our introductory definition, ecocomposition examines all sorts of spaces, including natural, urban, constructed, political, personal, virtual, and even imagined spaces. Second, ecocomposition, as McDowell suggests it might be, is not about the interpretation of environmental writing, be it in local newspapers or any literary text. Most recently, Terrell Dixon offered a brief definition of *ecocomposition* in the beginning of his essay "Inculcating Wildness: Ecocomposition, Nature Writing, and the Regreening of the American Suburb." He offers that ecocomposition classes are those "classes that emphasize reading and writing about nature and the environment" (77). Dixon's definition is a wonderful beginning point, and we will return to his definition and others' in chapter five, "Ecocomposition Pedagogy." For now we wish to borrow a portion of Dixon's definition and begin formulating our own by offering that ecocomposition is about the activity of writing, about the production of discourse. But Dixon's definition is only a small beginning. Let us explain in further detail.

Etymologically *ecocomposition* reflects *ecology*, a science that evolved specifically to study the relationships between organisms and their surrounding environment. Ernst Haeckel first defined "oecologie" in 1866 as "the total relations of the animal both to its organic and to its inorganic environment" and as "the study of all the complex interrelationships referred to by Darwin as the conditions of the struggle for existence" (quoted in Ricklefs, 1). Haeckel may as well have offered these words as the definition for contemporary composition studies. After all, composition studies in its post-cognitive, post-process, post-expressivist incarnation is also a study of relationships: relationships between individual writers and their surrounding environments, relationships between writers and texts, relationships between texts and culture, between ideology and discourse, between language and the world. Ecocomposition highlights the impact of the spaces in which discourse occurs, suggesting that most inquiries into these relationships do not fully account for the degree to which discourse is affected by the locations in which it originates and terminates. And, as we now discuss it, understanding these relationships is crucial to survival. Oppressive hegemonies manifest themselves in discourse; racial, cultural, sexist, classist oppression recurs through discourse. How we transgress those oppressive constructs, how we survive in them is a matter of discursive maneuvering. To paraphrase Haeckel, as Sid has explained in his essay "Writing Takes Place," ecocomposition is "the investigation of the total relations of discourse both to its organic and inorganic environment and the study of all of the complex interrelationships between the human activity of writing and all of the conditions of the struggle for existence." In other words, ecocomposition examines the relationships between discourse and environment. Ecocomposition inquires as to what effects discourse has in mapping, constructing, shaping, defining, and understanding nature, place, and environment; and, in turn, what effects nature, place, and environment have on discourse. As we have suggested and reiterate here because of the many misinterpretations of this aspect of ecocomposition, we mean *all environments:* classroom environments, political environments, electronic environments, ideological environments, historical environments, economic environments, natural environments. Ecocomposition must examine not only the relationships between discourse and "Nature," but the

relationships between discourse and any site where discourse exists. That is
to say that while ecocomposition has its roots in environmental ecology
and social ecology, it must also turn its inquiry to all of the sites in which
discourse is taught, studied, used, and lived. As Arlene Plevin writes, bor-
rowing from Norman McLean, ecocomposition "is more than smuggling
in an essay about trees—or even discussing the powerful pull of students'
favorite places. It is arguably a more radical move, one capable of continuing a postmodern teacher's desire to diffuse his or her authority—in decen-
tering the classroom. It is a move which is able to reduce, even critically
disrupt, the archetypal binaries of culture/nature, male/female, and even
human/non-human" (148). That is to say, the prefix "eco" must not be mis-
represented as simply "environmental" as it often is, but instead must be
understood specifically as a study of relationships. Ecocomposition is not
"writing about trees"; ecocomposition is the study of written discourse and
its relationships to the places in which it is situated and situates.

Ecocomposition locates writing in place and environment; it looks toward the ecology of language. Ecocomposition resists discursive maneuvers that create dualistic splits such as nature/culture and (hu)man-
made/natural; instead, ecocomposition argues for a more holistic approach
to seeing humans' place in the world. That is, ecocomposition contends
that identifying nature as an object separate from human culture and life
aligns it as an object that humans may act upon rather than within. This
particular discursive position (identifying nature as split from the rational
world of humans, and therefore subject to humans' domination) has its
roots in the scientific revolution's intellectual will to dominance. Many of
the contemporary discursive positions we will go on to critique have their
origins, if not their clearest statements, in sixteenth- and seventeenth-
century thinkers like Sir Francis Bacon and René Descartes. For Bacon, to
know nature meant to disturb and annoy it *(natura vexata),* and he argued
the anthropocentric view that "the whole world works together in the service of man; and there is nothing from which he does not derive use and
fruit" (quoted in Marshall, 184). For Descartes, the understanding and
control of nature are achieved by separation from the material world, followed by precise and careful measurement of it, in order to "make ourselves masters and possessors of nature" (quoted in Rifkin, 32). While these
positions sound extreme today, they capture a human-centered arrogance
still prevalent in many current discourses. Ecocomposition stresses a connected world view over separation of human life from nonhuman life and
biosystems, recognizing that such dualistic positions are and have been discursively constructed, and that more ecologically tuned perspectives are
only possible through more holistic discursive forms.

Coupled with this agenda, ecocomposition seeks to engender a critical awareness of how discourse creates natural places and how all environments affect written discourse. In other words, ecocomposition disagrees with ecocritic Harry Crockett when he writes "we resist the idea that reality is socially constructed." Ecocomposition identifies that all reality, including nature, is discursively constructed. "Environment" is (merely) an idea that is created through discourse. This is, of course, not to suggest that mountains, rivers, oceans, and trees do not actually exist. Such a suggestion would be pointless and unarguable. What we are suggesting, though, is that our only access to such things is through discourse, and that it is through language that we give these things or places meaning: historical, material, political, personal, natural, spiritual. For instance, Marine Life Conservation Districts in Hawaii, such as Kealakekua Bay (also known by its Westernized name, Captain Cook) prohibit fish collecting, fishing, and anchoring of boats. Violating these prohibitions results in severe penalties. In fact, conservation districts like this in Hawaii are often more fiercely protected by local residents than by federal authorities. A few miles north or south of Kealakekua Bay, however, fish collectors, fishermen, and boaters proliferate. It is doubtful that the fish or coral know that Kealakekua Bay has been named a Conservation District, yet in the eyes of most residents, the former is accorded almost sacred status while the later is seen as an economic commodity. There is no intrinsic difference between a Marine Conservation district and the rest of Hawaii's coastline; the difference is purely discursive. Similarly, the Florida State Park system has adopted the motto The Real Florida to identify natural Florida as opposed to developed Florida. The Real Florida is advertised on highway billboards and tourist brochures as the last bastions of natural Florida. The naming of certain fenced-off areas as "real" stands directly in opposition to all areas outside those fences which are "man-made." For example, large sections of the Everglades have been designated as nature preserves, and as such they are accorded special significance, with particular rules regarding how and when humans might interact with them. However, huge sugarcane fields lie just a few miles from many of these preserves, complete with sugar processing plants and distribution centers. There is (or was) no real distinction between the land within the preserves and the land that is used for sugar farming other than the distinctions that have been discursively accorded to them by humans. As evidenced by the many recent debates on land use in Florida, sugar farmers and preservationists have radically different definitions of what the Florida Everglades is or should be. So in a sense, there is no objective environment separate from the words we use to represent it. Like Carl G. Herndl and Stuart C. Brown, we argue that "the environment is not a thing you could go out and find in the world. Rather, it is a concept and an associated set of cultural values that we have constructed through the way we use language" (3).

That being said, however, we also note that ecocomposition addresses the current environmental crisis as a potentially catastrophic biospheric event that demands our consideration and action. Ecocomposition identifies the ecological relationships between humans and surrounding environments as dependent and symbiotic. It recognizes the decline of nature both discursively and materially. Like theorists Tom Jagtenberg and David McKie, we acknowledge the whole spectrum of our nonhuman physical environment as "so central to sustainable life that it undermines the very idea of space and the biophysical world as a context for human activity" (xii). While discourse does indeed shape our human conceptions of the world around us, discourse itself arises from a biosphere that sustains life. That is, while discourse "creates" the world in the human mind, the biospheric physical environment is the origin of life (and consequently, the human mind) itself. The relationship between discourse and environment is reciprocal. Similarly, the diversity and richness of language reflects the diversity of the world in which such language arises. For example, indigenous languages in two ecologically distinct locations, Alaska and Hawaii, reflect the geography and climate of each. The indigenous Hawaiian language has nearly as many terms and concepts (lexemes) to represent various forms of rainbows as the Eskimo language has for snow. In the Hawaiian language, we find the following terms for *rainbow:*

anuenue	rainbow
alaea	reddish rainbow
hakahakaea	greenish rainbow
onohi	rainbow fragment
uakoko	earth-clinging rainbow
kahili	standing rainbow shaft
luahoano	rainbow around sun or moon
punakea	barely visible rainbow

In Inuit, the best-known of the five Eskimo languages, we find the following terms for snow:

quanuk	snowflake
kanevvluk	fine snow particles
natquik	drifting snow particles
nevluk	clinging particles
aniu	fallen snow

muruaneq	soft, deep fallen snow
qetrar crust	on fallen snow
nutaryuk	fresh fallen snow

Our point here is not to count words, nor is it to suggest that one language is more complex than another. What we are suggesting is that our vocabulary says quite a lot about the particular environment in which we live. Language reflects place. So, in effect, preserving natural places, ecosystems, and the denizens of them is a move that preserves the fullness, depth, and precision of our discourse. As Nancy Lord (1997) writes in *Fish Camp: Life on an Alaskan Shore:*

> Languages, after all, belong to places in the same way that living creatures do. They're indigenous to the places that spawn them, both in the words needed to identify and address the particulars of that place and in the structure needed to survive there. Anyone who's ever studied a foreign language knows that even a modest familiarity with its vocabulary and grammar provides fascinating insights into the ways that a culture thinks about itself, what it values, and how it fits its origins. (58)

In a sense, humans occupy two spaces: a biosphere, consisting of the earth and its atmosphere, which supports our physical existence, and a semiosphere, consisting of discourse, which shapes our existence and allows us to make sense of it. We see these two central spheres of human life—the biosphere and the semiosphere—as mutually dependent upon one another. Where a healthy biosphere is one that supports a variety of simbiotic life forms, a healthy semiosphere is one that enables differences to coexist and be articulated. In both a material and a discursive sense, differences are a critical measure of a system's health.

Having said that, we want to extend this definition to expand upon how this inquiry of relationships is a multi-faceted area of study which draws on many other areas of inquiry including composition studies, feminism and ecofeminism, cultural studies, ecology, literary criticism and environmentalism. In the pages that follow, we provide a more intricate explication of *ecocomposition* by first providing a detailed overview of the most significant facets of ecocomposition to date. As with our initial definition, ecocomposition must fracture within itself as those working inside of its loose borders direct their attentions toward sub-specialties and disagree with one another over theory, method, and teaching. It must provide diverse approaches to theories and pedagogies. In the next five chapters, we offer these further impressions of ecocomposition as a starting place.

Natural Discourse

Having now introduced ecocomposition as a study of the relationship between discourse and environment and expressed an understanding that environment is as much a construct of discourse as discourse is a product of environment, it would seem odd that we would suggest in our title that discourse could be natural. That is, it would seem that we have been arguing that there is no such thing as Natural Discourse. However, we want to point out that our title is meant to suggest the deeply enmeshed relationship between ecological thinking and discourse studies and to question the very relationships between nature and discourse by highlighting the uncomfortable and problematic "naturalness" of discourse. In other words, our title aims to draw attention to the "where" and "how" of language use, foregrounding the fact that discourse always occurs within particular environments, that these environments are integral to the construction of language and knowledge, and that particular acts of communication have their own nature according to the circumstances and locations that precipitated them. Let us not forget, in fact, just how dangerous discourse can be when it is made to seem "natural." That is, oppressive discourses often maintain power when those discourses go unquestioned and are assumed to be an immanent and inherent part of things. We question not only the making natural of discourse, but also the very discourses that construct phenomena and objects as natural: natural discourse.

In the pages that follow, we continue our investigation into the relationships between environment and writing, place and discourse by first examining the academic and intellectual sites from which ecocomposition grows. Next, we consider the science of ecology and the evolution of ecological thinking and the ways that ecology might inform thinking in composition studies. We then explore the role of the activist intellectual and public writing in ecocomposition, inquiring into both the understanding of public spaces and the role of the intellectual in those spaces. Turning, then, from larger, public spaces, to classroom and pedagogical spaces, we analyze the ways in which ecocomposition pedagogies have evolved and the ways in which environmental issues have entered into composition classrooms. We offer approaches to developing ecocomposition pedagogies and strategies for ecocomposition classrooms. Finally, we turn to a study of the role of the personal in ecological thinking, critiquing the social constructionist assessment of the role of the individual, exploring the pathos of the ecology of writing, and ultimately turning to classical rhetoric to reconsider ecological thinking about discourse.

As we begin the rather large task of introducing ecocomposition, we take on a rather interesting project in research. This book only begins to scratch the surface of a body of research that needs to be further explored. In fact, in many ways, this book takes on the role of introduction in a rather encyclopedic fashion, exploring a wide range of materials, often only in brief acknowledgment rather than in depth. Of course, such a glossing of some research prohibits an exhaustive, comprehensive examination of any one aspect of ecocomposition. However, we have intentionally attempted here to cover as much introductory territory as possible to both introduce ecocomposition and to open as may avenues of travel within ecocomposition as possible. Our goal is to promote exploration and theorization in specific areas within and without ecocomposition, and we hope that others will find this introduction useful, drawing on what they find here and opening new paths in ecocomposition. In his book *The Gutenberg Elegies,* Sven Birkerts explores the effects of technology on the future of a culture of books, and in this sometimes troubling commentary on literature and the "impact of technology on reading" Birkerts argues that the move toward technological literacies and away from more traditional book literacies is conspicuously marked by what he refers to as "the gradual displacement of the vertical by the horizontal—the sacrifice of depth to lateral range," or as he explains it, "a shift from intensive to extensive reading" (72). What Birkerts argues is that when books were more rare, when texts were not easily available, readers spent more time digging for depths of meaning with each text. He argues that "in our culture, access is not a problem, but proliferation is" (72). Hence, he contends that contemporary readers tend "to move across surfaces, skimming, hastening from one site to the next without allowing the words to resonate inwardly. The inscription is light but covers vast territories" (72). Though we have some problems with Birkerts' dismissal of contemporary readers and reading an act in which he claims "quantity is elevated over quality"—we do wish to borrow his metaphor of horizontal and vertical. This book is, unquestionably a horizontal study. It traverses wide territory, ranging from the history of ecology to public intellectualism to composition pedagogy to rhetoric and a host of other subjects. But this book crosses these territories to begin to draw early maps, to locate those very sites that demand further vertical inquiry. We envision *Natural Discourse: Toward Ecocomposition* specifically as a move toward ecocomposition research, as an introduction. We hope that as you read these introductory moves, you note locations in need of further study, sites ripe for research. As we mentioned earlier, this book is a first foray into defining ecocomposition as an inquiry; it is by no means an end.

CHAPTER 2

The Evolution of Ecocomposition

꣸꣯꣭

> *How much finer things are in composition than alone.*
> —Ralph Waldo Emerson, 13 July 1833,
> *The Journals of Ralph Waldo Emerson*

Many in composition have traced the discipline's history. In fact, most introductory graduate courses in composition begin by reading North, Miller, Berlin, or the likes. One of the more recent and more important developments in composition's history has been a move post-process. As Gail Hawisher, Paul LeBlanc, Charles Moran, and Cynthia L. Selfe explain, "During the period of 1983–1985, composition studies absorbed the changes brought about by the new emphasis upon process and began to chart the course it would follow post-process, looking beyond the individual writer toward the larger systems of which the writer was a part" (65). In other words, post-process composition studies began to focus on issues of social construction rather than on issues of the individual writer working within an individual process. Composition began to examine the environments in which writers write under the rubrics of culture, class, gender, race, and identity. Identifying these larger influential "systems," as Hawisher et al. label them, afforded teachers of writing the opportunities to teach definable, codifiable systems as "conceptual schemes" (Donald Davidson's phrase) that dominate discourse production. That is, composition turned toward issues of social construction, of race, of gender, of culture, of identity, in order to better understand both how external social constructs affect writers, and how, in turn, writers impact those same discursively constructed critical categories. In essence, composition did not just make a move post-process, composition made a move into ecology.

Ecology evolved as a science that sought to explore the relationships between organisms and their surrounding environments. In turn, ecology also began to study ways in which individual organisms and populations affected those same environments. Ecology, however, is also rooted in notions of management; that is, early ecology sought to understand these relationships in order to better exploit nature for human consumption. In fact, Ernst Haekel, who first defined *ecology,* was heavily influenced by the writings of Karl Marx, and those influences can be seen clearly in ecology's reliance on cost/benefit and other economic metaphors. Ecology, then, evolved as a science of natural management as much as a science of relationships. Composition has a similar agenda, and the influences of Marxism on the discipline can clearly be seen in discussions of liberatory learning and radical pedagogy, in conversations about class, and in research regarding educational systems. In many instances, composition now explores the relationships between individual writers (identity) and local environments (ideology, space) as well as ways in which populations interact with environment (culture). Environment is studied to see what impact it has on writers under the headings of *ideology, dominant discourse, hegemony,* and the likes. And, in turn, composition has stressed that such information can provide access to understanding how to better manipulate and manage discursive situations. Just as liberatory pedagogies, for instance, suggest ways in which marginalized borderland voices may be empowered in hegemonic discourse—the very environment that determines what is borderland and what is a marginalized voice—ecology sought to find better ways for humans to empower themselves over local environments. Similarly, as much as composition is a discipline that is enmeshed with issues of consumption and production, so too is ecology. Ecologists, for instance, talk about the "economy of nature" and examine the consumption and production of energy. In fact, compositionists talk about the consumption and production of discourse in much the same way ecologists discuss the consumption and production of energy.

As we mentioned, the mid-1980s saw composition move from examining individual writer's cognitive processes to inquiries regarding the interactions between writers and social forces which acted upon writers and upon which writers had effect. Marilyn Cooper, in support of this move post-process, contends that developing pedagogies that address more social issues of writing "signals a growing awareness that language and texts are not simply the means by which individuals discover and communicate information, but are essentially social activities, dependent upon social structures and processes not only in their interpretive but also in their constructive phases" (5). Cooper's 1986 article "The Ecology of Writing"—which first appeared in *College English* and later in the collection *Writing as Social Action*—was central to the post-process move. In this article, Cooper proposes an ecological model of writing, "whose fundamental tenet is that writing is an activity through which a person is continuously engaged with a variety of socially constituted systems" (6).[1] That is, Cooper suggests that writing not be examined as individual process, but as an entity reliant upon environment and, in turn, a force on that very environment. Cooper, however, does not use the term *environment;* rather she identifies *systems* that affect and are affected by writers. Cooper's ecological model is one of the first moves into ecocomposition. We wish to extend Cooper's model to identify that all environments play the same role as the systems that Cooper identifies:

> An ecologist explores how writers interact to form systems: all the characteristics of any individual writer or piece of writing both determine and are determined by the characteristics of all other writers and writing systems. An important characteristic of ecological systems is that they are inherently dynamic; though their structures and contents can be specified at a given moment, in real time they are constantly changing, limited only by parameters that they themselves subject to change over longer spans of time. (6–7)

With this definition, Cooper begins the formal study of ecocomposition, though at the time the term had not been developed. However, Cooper limits her definition to writers' effects on writers, and ecocomposition must take into account other environmental effects: history, gender, culture, race, class, and—literally—local environments. Granted, each of these categories—environment included—should be identified as socially constructed categories. In other words, each of these categories is indeed affected by writers; each is created/constructed by writers.

What is critical about Cooper's ecological model is that it introduces composition to the notion that writers interact with systems that affect their writing. As Erika Lindemann explains when writing about Cooper's article and about writing as system, "The ecological model usefully complicates the learning and teaching of writing because it reminds us of the social context in which all writers work" (1995). Writers, in essence, are organisms dependent upon their surroundings—surroundings that are dynamic, difficult to define, and susceptible to the forces imposed by writers. As Cooper notes, ecological models are not simply new ways of saying "contextual" (6). Context suggests that potential effects of *all* local systems can be identified through heuristics in order to provide writers with accurate, and complete, information prior to writing. Cooper points out that "in place of the static and limited categories of contextual models, the ecological model postulates dynamic interlocking systems that structure the social activity of writing" (7). She continues, "The systems are not given, not limitations on writers; instead they are made and remade by writers in the act of writing" (7).

Cooper explains that "the metaphor for writing suggested by the ecological model is that of a web, in which anything that affects one strand of the web vibrates throughout the whole" (9).

> Two determinants of the nature of a writer's interactions with others are intimacy, a measure of closeness based on any similarity seen to be relevant—kinship, religion, occupation; and power, a measure of the degree to which a writer can control the action of others. . . . Writers may play a number of different roles in relation to one another: editor, co-writer, or addressee, for instance. Writers signal how they view their relationship with other writers through conventional forms and strategies, but they can also change their relationships—or even initiate or terminate relationships—through the use of these conventions if others accept the new relationship that is implied. (8)

Cooper's model is Darwinian in that it suggests power relationships similar to species survival.[2] Writers seek voice, agency, and empowerment to ensure discursive (and sometimes literal) survival. While many would say it is abhorrent to compare human individual's fight for discursive survival with species survival, ecocomposition recognizes both as important, neither as more critical, and the very claim that human life takes precedent as an anthropomorphic, homocentric, ecocolonial act of Darwinianism in and of itself—an oppressive maneuver. As we will explain later in this chapter, a basic tenet of ecocomposition is the resistence of hegemonies which oppress nature in ways similar to the oppression of women and other colonized groups.

Cooper concludes her argument against cognitive models by identifying that "In contrast, then, to the solitary author projected by the cognitive model, the ideal image the ecological model projects is of an infinitely extended group of people who interact through writing, who are connected by the various systems that constitute the activity of writing" (12). We would like to extend Cooper's statement to include not just writers as affecting writers, but larger environments; that is, ecocomposition, as we see it, looks beyond writers and readers to larger webs. But, as Cooper points out:

> It is important to remember that the image the ecological model projects is again an ideal one. In reality, these systems are often resistant to change and

not easily accessible. Whenever ideas are seen as commodities they are not shared; whenever individual and group purposes cannot be negotiated someone is shut out; difference in status, or power, or intimacy curtail interpersonal interactions; cultural institutions and attitudes discourage writing as often as they encourage it; textual forms are just as easily used as barriers to discourse as they are used as means of discourse. A further value of the ecological model is that it can be used to diagnose and analyze such situations, and it encourages us to direct our corrective energies away from the characteristics of the individual writer toward imbalances in social systems that prevent good writing. (12–13)

This is particularly true as we engage ecocomposition as a method for introducing students to various writing scenarios in various environments. Quite frequently, when students or other writers enter into a web, it is difficult for them to create enough disturbance to shake the web. Instead, environments often subsume organisms into the ecosystems to maintain operational integrity. Approaching hegemony as an environmental structure can shed a good deal of light on ways in which environments can maintain particular relationships in order to retain its structure. Cooper's web provides an interesting metaphor for examining hegemony.

Likewise, Richard M. Coe, in his award winning article "Defining Rhetoric—and Us: A Meditation on Burke's Definitions," also began to question the relationships between writers and surrounding environment. In his examination of how Kenneth Burke defines humans as human, Coe specifically addresses the nature/culture split in order to examine how rhetoric and language have been used to define the human species as separate from nature, yet functioning in relationship to it. He writes, "we are removed from nature" (44). For Coe this is a critical point, since his argument is that Burke's definitions of what makes us human offers important lessons for compositionists, particularly as compositionists attempt to study the role of language and culture. That is to say, Coe turns to Burke in order to begin to develop an ecological methodology for exploring the ways in which discourse helps shape the very relationship between humans, culture, and nature. Coe's work, like Cooper's, initiates ecocomposition by introducing concepts of human discursive participation in larger systems, including those often labeled as "natural."

For Coe it is critical that as we examine the work of Burke that we recognize that "Burke begins (and ends) with a definition of *humanity*" (39). He goes on to say that "any comments on matters cultural, Burke asserts, must embody assumptions about the nature of human beings who compose culture" (39). Here, we read Coe, and in turn Burke, to be referring not to some inherent condition of humanness—"the nature of human beings"—while composing culture, but instead read "nature" to refer to the position of human beings within larger systems. "The nature of human beings," then, is the place where humans fit in relation to other organisms—human and other—and environments while they compose culture. And, as Burke uses the word "animal" while defining humans (in part) as "the symbol-making, symbol-using, symbol-misusing animal," it seems likely that Burke, and Coe, are rather conscious of the use of *nature* and *animal* in larger ecological senses. To quote Coe at length:

> The *genus* of Burke's definition, "animal," is both true and false. We are, biologically, animals; but we are defined, distinguished from other animals, by our use of symbols (instruments of our own making), especially language (the tool that is more than a tool), which allows us to develop a culture, to think abstractly and morally about our experience past, present, and future. Our culture separates us from nature, creates the nature/cultural boundary. It frees us—but in the process alienates us (from our natural condition). Our condition becomes more social than natural, shaped by culture within only very broad biological and ecological parameters.
>
> Culture in this sense *negates* nature, though *negate* must be understood dialectically, for nature is not destroyed by our transcendence, and we remain in nature as we go beyond it. *Beyond* is a key word for Burke, who talks about "beyonding": (in part to evade the technical philosophical term, *subulate*). Thus, separation does not mean we are not connected, just that a boundary has been drawn (hence the need for connection). The social is, in at least one crucial sense, natural, derived from nature, from the evolutionary process. We are "bodies that learn language/thereby becoming wordlings." (45)

Coe goes on to examine Burke's definition of what makes us human, never leaving the idea that the very definitions he and Burke are working with are the very things that separate humans from nature, for "we conceive a potential and strive to actualize it." Coe continues:

> But this separates us from nature, from the here-and-now, makes us part of the fallen world, where things die and rot, leads us into all kinds of confusions and misapprehensions, this rotting within us, this dissatisfaction with what is, this humanness, our downfall and our wonder, our specialness, our potential to be more than what we are. (46)

Of course, ultimately, the payoff for ecocompositionists in this discussion is Coe's conception of rhetoric as the very thing that defines humanity:

"what makes us human is our culture, which is founded in our unique form of symbolizing, our language, which is in its very nature rhetorical as it goads/gods us, moves/motivates us, makes us social, cultural (non-)animals, allows us to compose ourselves humanly. The study of language, culture, discourse, rhetoric, and humanity is one" (46). As Coe puts it, "for Burke *Homo sapiens* is synonymous with *Homo rhetorica;* human wisdom is a discursive practice" (46).

There is little question that Coe's project is one of ecocomposition, and it raises important questions for ecocompositionists to consider. For just as we have argued that ecocomposition resists the culture/nature binary, so too does ecocomposition recognize that such a split has been a predominant aspect of Western thinking. Ecocomposition, in line with Coe's work, seeks to examine how and why humans conceive of themselves as separate from nature and the ramifications this conceptualization has on both discourse and environment/nature. Similarly, Coe's notion that "The social is, in at least one crucial sense, natural, derived from nature, from the evolutionary process" (45) is critical to both ecocomposition's understanding of the relationship between nature and culture and ecocomposition's intersections with cultural studies (which we explore in detail later in this chapter). That is to say, Coe's project here also contributes to the beginning of ecocomposition in landmark ways.

While composition has been slow to turn to Coe's use of Burke, ecocomposition should acknowledge and embrace this work. Of course, Marilyn Cooper's web and Cooper's ecological model, like much of composition's move post-process, has been critical in examining writers' relationships with external forces, and thankfully, composition has embraced the turn toward such examinations with enough vigor that such inquiries now dominate composition studies. That is to say, since the mid-1980s, composition inquiry has, in many ways, become an ecological study. In fact, if compositionists are to understand composition studies as being more than simply a pedagogical endeavor and instead identify it as a discipline which examines discourse and its relationships to writing, culture, gender, environment—in essence, to everything—then compositionists must identify that what we really do is study relationships between discourses and environment, discourses and writers, discourses and other discourse.

In other words, the intellectual activity of composition studies is already an ecological endeavor. It is a project which examines the relationships between discourse and environments, including classroom environments, electronic environments, public spaces, and so on. Perhaps identifying composition as the study of discourse with a tendency toward scientific methodologies may seem offensive to some. Many may argue that composition studies is specifically humanist and that it must resist the reductionary processes of sciences. But, ecocomposition and composition are integrative studies, studies which should turn to other bodies of knowledge, studies which should be interdisicplinary. For example, Greg Myers mentions ecology in his analysis of the social construction of two biologists' proposals. Myers suggests a close relationship between ecology, the science of natural environments and the relationships between its inhabitants, and social constructionist approaches to composition, which see writing as an activity that is negotiated by individuals in discrete locations. Myers argues that there is much that compositionists can learn from the study of ecology, and he concludes by observing that "we should not only observe and categorize the behavior of individuals, we should also consider the evolution of this behavior in its ecological context" (240). Ecocomposition, in particular, must rely not only on ecology, but on a host of other theoretical and pedagogical schools of thought from which it borrows and contributes.

Ecocriticism

Ecocomposition grows from and turns to ecocriticism in its development; yet ecocomposition is not an extension of ecocriticism. Several reviewers of our collection *Ecocomposition: Theoretical and Pedagogical Approaches* had been insistent that when we discuss ecocomposition we do so as a subdiscipline of ecocriticism. It is not. Ecocomposition borrows from ecocriticism, but grows on its own. Ecocriticism is a literary criticism that looks toward textual interpretation; eocomposition works from the same place, but is concerned with textual production and the environments that affect and are affected by the production of discourse. We have been grateful for the work being done by scholars affiliated with the Association for the Study of Literature and the Environment (ASLE) and ASLE's encouragement of the development ecocomposition. In addition, as we have noted, the move to include ASLE in CCCC with a redirected focus toward composition has been of great benefit to the development of ecocomposition, including Randall Roorda's efforts in establishing the ASLE Special Interest Group at CCCC. However, we are also cautious of the evolutionary grounding of ecocomposition in ecocriticism and move to step beyond ASLE and its literary focus. That is, we are cautious that ecocomposition direct its focus not on literary criticism or even textual interpretation in the larger sense, but that it evolve as its own inquiry into the relationship between the production of written discourse and environment.

Ecocriticism, though coined by William Rueckert in his essay "Literature and Ecology: An Experiment in Ecocriticism," was introduced to American literary scholars by Cheryll Glotfelty at the Western Literature Association (WLA) meeting in 1989. Glotfelty urged the adoption of the term as a critical approach to studying nature writing. By 1994, the term was widely used, though participants at WLA were still unsure of its exact definition. Since that time, many have tried to come to agreement as to what ecocriticism is. In the 1996 collection *The Ecocriticism Reader,* Glotfelty posits that

> Simply put, ecocriticism is the study of the relationship between literature and the physical environment. Just as feminist criticism examines language and literature from a gender-conscious perspective, and Marxist criticism brings an awareness of modes of production and economic class to its readings of texts, ecocriticism takes an earth-centered approach to literary studies. (xviii)

Glotfelty continues to note that ecocritics ask a range of questions such as "How is nature represented in this sonnet? What role does the physical setting play in the plot of this novel? Are the values expressed in this play consistent with ecological wisdom? How do our metaphors of the land influence the way we treat it? How can we characterize nature writing as a genre?" (xix). More theoretically, ecocritics might also ask "Do men write about nature differently than women do? In what ways has literacy itself affected humankind's relationship to the natural world?" (xix). Most important, however, ecocriticism provides two critical components to ecocomposition: first, the inquiry as to whether "in addition to race, class, and gender, should place become a new critical category?" (xix). And second, the notion that "all ecological criticism shares the fundamental premise that human culture is connected to the physical world, affecting it and affected by it"(xi).

The questions regarding nature's representation in texts is important to ecocomposition; however, as Glotfelty poses her sample questions, these questions are too confining and restrictive for ecocomposition. The critical goals of ecocriticsm limit ecological understandings of writing to inquiries pertaining specifically to writing about nature. Ecocomposition would like to ask questions such as "what effects do local environments have on any kind of writing, any kind of writer?" This question must include asking about the political environments of classrooms, the technological environments of cyberspace, the very metaphors by which we define writing spaces, not only natural places about which writers write. We must ask not just do men and women write about nature differently, but do environments affect the difference between men and women writers? Ecocomposition conforms with Glotfelty's premise that "if we agree with Barry Commoner's first law of ecology, 'Everything is connected to everything else,' we must conclude that literature does not float above the material world in some aesthetic ether, but, rather, plays a part in an immensely complex global system, in which energy, matter, and ideas interact" (xix). Commoner's law is reflected in Cooper's web. However, ecocomposition is less likely to turn to the field of study "literature" and critical interpretation, but instead look toward text, toward discourse in more encompassing ways and claim that language does not exist outside of nature. Ecocomposition's focus on discourse takes in more than just textual interpretation; it looks at discourse as the most powerful, indeed, perhaps the only, tool for social and political change. We agree with Barry Lopez when he argues that "this area of writing [environmental writing] will not only one day produce a major and lasting body of American literature, but that it might also provide the foundation for a reorganization of American political thought" (297). Lopez is correct that the activity of writing, the production of environmentally conscious writing can affect change, not the critical analysis of that body of writing. And despite claims by ecocritics such as Thomas K. Dean, Harry Crockett, Christopher Cokinos, and a host of others, that ecocriticsm seeks to change environmental policy and "green" the world, the activity of literary criticism—à la Stanley Fish— does not affect change; activism and action affects change. Teaching students to resist hegemonic discourses that create anti-environmental legislation affects change. Reading Edward Abbey does not. As Raul Sanchez has commented in an on-line conversation regarding the radical and political possibilities of interpretation/reading, "I'm always puzzled at the implicit and explicit faith some seem to have in the power of hermeneutics as if celebration or even proper postcolonial 'interrogation' of literary texts could represent anything remotely like radical political action." Sanchez later comments that he finds the "more useful" activity of teaching writing to lend toward political action. Likewise, compositionist Susan Miller has noted quite accurately that textual interpretation (reading) is not writing, as she addresses the difference in power between interpretation and production (Technologies, 499). Encouraging students to be critical of the very environments in which they produce discourse and the effects those environments have upon their writing affects change. Hence, ecocomposition's split with ecocriticsm comes from the will to examine the activity of textual production rather than to engage in textual interpretation. Simply put, if ecocriticism looks toward textual interpretation, ecocomposition is interested in examining the *activity* and *locations* of textual production as well as all of the other environments which affect and are affected by the production of discourse.

We are particularly concerned with the link between ecocriticism and ecocomposition as it has been manifest in recent discussions of ecocomposition. Similar to CCCC's acceptance of ecocomposition through ASLE, much of the current pedagogical work being generated in ecocomposition stresses the use of texts that can easily be categorized as "nature writing." For instance, many of the textbooks currently on the market that intend to encourage students to think more ecologically and develop environmental awareness address nature writing texts as ripe for criticism and analysis. In other words, they advocate the analysis of literature. Such an approach to composition offers no real insights into the activity of writing and does little to extend students' conceptions of how their discourse is already ecological. Many of the readers that are designed for composition classrooms such as Scott H. Slovic and Terrell F. Dixon's *Being in the World: An Environmental Reader for Writers,* Richard Jenseth and Edward E. Lotto's *Constructing Nature: Readings from the American Experience,* Carolyn Ross' *Writing Nature: An Ecological Reader for Writers,* and even Sid's new text *Saving Place: An Ecoreader for the Composition Classroom,* to name a few, are designed to provide readings for student writers. That is, environmental issues are presented as subjects which students read and write *about,* think *about,* talk *about* rather than participate *in.* That is not to say the editors of these collections nor those writing about ecological pedagogies for the composition classroom do not take hands-on experience as a primary agenda in their pedagogies. However, what we hope for ecocomposition is that it not only address environment as merely another subject students may write about, but also rather a critical instrument for understanding the very function and operation of writing. In other words, while examining nature writing certainly begins to forward the basic ecocomposition project, until students and teacher/scholars move beyond this limiting view of ecocomposition, ecocomposition will simply be identified as another flashy subject about which teachers can assign writing, but that really does not shed any light on the activity of writing. Ecocomposition is also about the environment of writing, and must step beyond ecocriticism's agenda of textual interpretation. (See David Sumner, "Don't Forget to Argue.")

Ecocomposition and Cultural Studies

It would seem, at first, that ecocomposition might be defined as a subdiscipline of cultural studies. After all, many ecocritics define part of the agenda of ecocriticism as making the critical category of environment/place as important in literary analysis as are culture, race, class, and gender, and many in cultural studies have begun to consider issues of space and location in their inquiries. Likewise, because of the close links between ecocomposition and ecocriticism, many have also begun to see ecocomposition as a means to introduce place and environment as a critical category and as a subject of study much like race, class, gender, and culture. Just as ecocriticism has certainly aided in the development of ecocomposition, some cultural studies work does engage issues critical to ecocomposition. Yet, the agenda of ecocomposition is to move beyond locating place and space as only a critical component of other larger inquiries and look to environment as the central issue in understanding self, identity, and existence. In other words, one goal of ecocomposition is the eventual development of epistemological acceptance of human's interconnectedness with nature and environment, not simply the use of these as categories for inquiry. At times, ecocomposition may seem to blend with cultural studies, ecocriticism, and other areas of study, yet it must move beyond the sub-facet of cultural studies. Cultural studies often investigates and analyzes the ways in which social forces and practices may construct a particular environment, environmental issue, or environmental moment such as a geographical, historical, economic, technological, racial, or other issue as pertaining to environment. Some cultural studies theorists have, in fact, noted the effects of environmental degradation on the construction of identity and culture. Stuart Hall, for instance, contends that "at one and the same time people feel part of the world and part of their village. They have neighborhood identities and they are citizens of the world. Their bodies are endangered by Chernobyl, which didn't knock on the door and say 'Can I float radiation over your sovereign territory?' Or another example, we had the warmest winter I've ever experienced in England last year—the consequence in part of the destruction of rain forests thousands of miles away" (343). Hall is among the first to argue for more ecological understandings of culture and identity, and such work is clearly aligned with the work of ecocomposition. While most cultural studies investigations have overlooked the degree to which environment impacts culture, Hall's work promises to reinvigorate cultural studies as a more ecological field of inquiry. Hall argues: "An ecological understanding of the world challenges the notion that the nation-state and the boundaries of sovereignty will keep things stable because they won't. The universe is coming!" (343).

Most current cultural studies projects have their theoretical origins in the work of British cultural theorists in the 1950s, particularly the work of Raymond Williams and Richard Hoggart. Williams and Hoggart saw cultural studies as the "entire lived experience of human agents in response to their concrete historical conditions" (Berlin and Vivion, viii). Moreover, they argued that these responses could not be reduced to merely economic or political bases, as Marxist conceptions of culture had suggested. In short, they conceived of culture as arising from particular ecologies that included, but could not be reduced to, the realms of the political and the economic. Much of the work in cultural studies since the fifties has incorporated discourse as an integral aspect of a culture's ecology, suggesting that language and other "signifying practices" are involved in the shaping of consciousness. However, while such cultural-studies investigations are critical to understanding the relations between discourse and environment, such investigations do not always address the issue of writing per se; issues of textual production are fundamental to ecocomposition, not to cultural studies.

There is a great deal of difficulty in exploring the relationship between cultural studies and ecocomposition. First, because both ecocomposition and cultural studies are multi-faceted endeavors, there is certainly room for parts of ecocomposition to intersect with parts of cultural studies inquiries, just as culture and environment, culture and nature are co-constitutive. For instance, we have argued that nature is itself a cultural construct. Similarly, we also assert that cultural studies is dependent upon environment and that any examination of culture must take into account environment and place as well. Of course, this is not a new claim as many engaged in the cultural studies project have identified place, terrain, weather, architecture, and space as key components of the construction of identity, of social relations, of economic relations, of human emotion, and of power. The mapping of particular places, ideas, or features is a recurrent theme and metaphor in recent cultural studies. This interest in mapping and space has arisen as a corrective to many previous cultural and social analyses, which privileged historical (time-based) investigations over locationary (space-based) investigations. As Tom Jagtenberg and David McKie suggest, "Texts, discourses, language, and culture can generally all be construed as maps by which we find our way around in life and through which social power is encoded regulated and reproduced" (1997, 40). Fredric Jameson's use of the term *cognitive mapping* is perhaps the clearest example of space-oriented cultural studies. In Jameson's "Cognitive Mapping" (1988), he introduces the term as a "spatial analysis" of culture (348). He goes on to suggest a concept that is akin to much of the recent work in composition studies, for example, mapping strategies emerge from different perspectives, and these particular perspectives play an integral role in the framing of discourse and taken-for-granted realities.

The concept of cognitive mapping runs through much of Jameson's work. For instance, his discussion of the urban hotel and the experience of the height, the glass, the confusion is clearly an example of spatial analysis. His questioning of which is out and which is in and the quintessential postmodern experience certainly falls within the inquiry of ecocomposition in that Jameson specifically asks as to how these spaces are produced, prior to how they are interpreted. Again, for ecocompositionists, the question of production is primary. Similarly, in "Postmodernism, or The Cultural Logic of Late Capitalism," Jameson contends that human culture has subsumed Nature and constructed it as artificial. That is, according to Jameson, culture has reached the "moment of a radical eclipse of Nature itself" (71). Jameson argues that culture must move toward a "practical reconquest of a sense of place," which he offers occurs through the metaphoric practice of "cognitive mapping." That is, for Jameson, we must establish "an *imaginary* relation to the *real*." We understand this relationship to suggest that humans need to relearn to imagine place as "real," as constructions that affect cultural identity.

Other recent investigations in cultural studies emphasize the mapping of various spaces as referential to nearly every aspect of human life. Jody Berland's "Angels Dancing: Cultural Technologies and the Production of Space," for instance, explores the spatiality of music. Berland suggests that "in theoretical terms, we need to situate cultural forms within the production and reproduction of capitalist spaciality" (39). She goes on to explore capitalism as both an architect and a consumer of space, arguing that it "builds a physical landscape appropriate to its own condition at a particular moment in time, only to have to destroy it, usually in the course of crises, at a subsequent point in time" (45). Similarly, Patricia Yaeger's *Geography of Identity* and Derek Wright's essay "Parenting the Nation: Some Observations on Nuruddin Farah's *Maps,*" in *Order and Partialities* also blend cultural studies and postcolonial studies concerns with inquiries of importance to ecocomposition.

While notions of spatiality and mapping have become an important aspect of cultural studies, many of its foremost scholars—ranging from Stuart Hall to Simon During to Larry Grossberg—insist that cultural studies is not any one thing, does not commit to any one methodology, and that it defies categorization and definition. Tony Bennet argues that cultural studies is "a term of convenience for a fairly dispersed array of theoretical and political positions which, however widely divergent they might be in other respects, share a commitment to examining cultural practices from the point of view of their interaction with, and within, relations of power" (23). Cary Nelson contends that cultural studies is an interdisciplinary, transdisciplinary, and sometimes counterdisciplinary field that operates in the tension between its tendencies to embrace both a broad anthropological and a more narrowly humanistic conception of culture. Unlike traditional anthropology, however, it has grown out of analyses of modern industrial societies. It is typically interpretive and evaluative in its methodologies, but unlike traditional humanism it rejects the exclusive equation of culture with high culture and argues that all forms of cultural production need to be studied in relation to other cultural practices and to social and historical structures. Cultural studies is thus committed to the study of the entire range of a society's arts, beliefs, institutions, and communicative practices. (4)

He goes on to say that "continuing preoccupation within [cultural studies] is the notion of radical social and cultural transformation and how to study it. Yet in virtually all traditions of cultural studies, its practitioners see cultural studies not simply as a chronicle of cultural change but as an intervention in it, and see themselves not simply as scholars providing an account but as politically engaged participants" (5). Likewise, Graeme Turner writes that cultural studies "defines itself in part through its disruption of the boundaries between disciplines, and through its ability to explode the category of 'the natural'—revealing the history behind those social relations we see as the products of a neutral evolutionary process" (6).

Because ecocomposition examines the relationships between discourse and environment, and since environment subsumes culture, ecocomposition might easily be classified as a branch of cultural studies simply because methodologies are similar. Certainly, Nelson's definitions require an understanding of relationships even though he is primarily focused on human life and culture. Composition's contribution to cultural studies has been to identify the critical categories subsumed under the heading cultural studies—race, class, culture, and gender—as discursive constructs. That is, composition has argued, in part, that the very ideas of race, class, gender, and culture are constructed through human interaction, conversation, and consensus. In many instances, identifying these categories as discursive constructs has provided the necessary access and metaphors that allows the same categories to be seen for what they are: ideologically oppressive constructs which are often used to support hegemonies.

Ecocomposition seeks to elicit inquiry into the role of place in the construction of identity. One of the most important facets of ecocomposition is the argument that while place and environment are also socially and discursively constructed, this category is of equal—if not greater—importance in the construction of identity. Ecocomposition contends that the place in which discourse emerges directly affects that very discourse. That is, ecocomposition posits that environment precedes race, class, gender, and culture. Of course, as we have argued, that very environment is known only through discourse. Hence, ecocomposition gives way to a theoretical chicken and egg argument: discourse creates environment and environment creates discourse. It is a dialogic relationship; environment and discourse are co-constitutive. This close enmeshing between discourse and environment necessitates that prior to examining the relationship between discourse and culture, race, class, or gender, environment must be considered. That is, while cultures certainly define environments, the places from which cultures evolve define those very cultures. That is, like Coe, we argue that the social, cultural, individual is dependent upon the natural and constructed environments from which they grow. In many ways, environment precedes culture. Of course, many humanist scholars will be outraged at the suggestion that these critical categories take a back seat to ecological/environmental concerns; after all, the human condition and issues of oppression are critical to liberation and human existence. Yet, as we see it, this is precisely why ecocomposition must move beyond cultural studies and why we must begin our inquiry into identity and cultural studies with an inquiry into our ecological relationships with the places from which our cultures emerged. What ecocomposition claims is that every living organism and every aspect of every environment is in some degree imbricated with the development and continuation of culture. That is not to say, however, that by examining environment we can come to some truer or more authentic understanding of culture or identity. We only mean to suggest that by examining environment, we might develop a clearer picture of our investigations of culture.

One of the major flaws of some "cultural studies projects" has been their focus on human life, first, as more important than all other forms of life, and, second, as separate from life, from the places where life happens. Cultural studies is mostly a homocentric project. Much like ecofeminism (which we will discuss in a moment), ecocomposition rejects the culture/nature split which some cultural studies seems to promote. Ecocomposition embraces William Howarth's notion that "although we cast nature and culture as opposites, in fact they constantly mingle, like water and soil flowing in a stream" (69). That is to say, ecocomposition locates human culture within environment, not as the defining agent of environment. Here lies the chicken and egg, and further reason why ecocomposition cannot be part of cultural studies. When ecofeminists such as Greta Gaard or Janet Biehl talk about "locating humans and animals within nature," they do so under an ecofeminist ideology that sees Nature as an encompassing

"Natural" system within which humans are a "Natural" part. Nature is Nature (capital *N*). However, ecocomposition recognizes that nature is a discursive construct while not relinquishing the biological connections humans have to Nature. That is, ecocomposition contends that access to environment is only provided through discourse, and though nature can only be known through discourse, Nature is a preexisting system/condition in which the human animal is only a part. All life forms come from Nature; all discourse comes from life forms. Nature produces all discourse. Respiration, for example, is a necessary precondition to discourse. Obviously, individuals can't think, speak, write, or even exist outside of the particular sustainable biosphere we call Earth (unless they work for NASA). However, without discourse, those same individuals can't organize the experience of being a breathing entity in a biosphere named Earth. This is why ecocomposition is not cultural studies: culture, race, class, and gender are certainly discursive constructs like environment and nature, but they are not preexisting Natural occurrences. That is, there is no Race with a capital R, no Culture with a capital C, no Gender with a capital G, and no Class with a capital C; all are solely human constructs defined through discourse, mediated through each other, and re-inscribed through ideology. To suggest that Race, Class, Culture, or Gender existed prior to discursive production would be to argue for racism and oppression as Natural. Ecocomposition resists such claims adamantly. To argue for Nature as a preexisting condition of human existence, however, is to identify the human animal as part of larger biological/ecological system and to identify that humans only have access to those systems through discursive naming. This is why studies of ecofeminism, ecoracism, and ecocolonialism are as critical in the evolution of ecocomposition as are cultural studies and ecocoriticism.

Ecofeminism

One of the most important theoretical endeavors from which ecocomposition borrows is the work being done in ecofeminism. Though the term *ecofeminism* was first used by Francoise d'Eaubonne in 1974 as a political call to action, like ecocomposition, ecofeminism is a multifaceted area of study which is difficult to limit to just one definition. Karen J. Warren offers that "just as there is no one feminism, there is neither one ecofeminism nor one feminist philosophy" (1996, x). Ecofeminism, as Greta Gaard explains, "is a theory that has evolved from various fields of feminist inquiry and activism: peace movements, labor movements, women's health care, and the anti-nuclear, environmental, and animal liberation movements" (1). Warren contends that with the rise of interest in women's movements and environmental movements, "the goals of these two movements are mutually reinforcing and ultimately involve the development of world views and practices which are not based on models of domination" (1996, ix). Gaard explains that "Drawing on the insights of ecology, feminism, and socialism, ecofeminism's basic premise is that the ideology which authorizes oppressions such as those based on race, class, gender, sexuality, physical abilities, and species is the same ideology which sanctions the oppression of nature" (1). At its core, ecofeminism seeks to end all oppression and recognizes that any attempt to liberate any oppressed group—particularly women—can only be successful with an equal attempt to liberate nature. This recognition is notably identified by Alice Walker who writes: "Some of us have become used to thinking that woman is the nigger of the world, that a person of color is the nigger of the world, that a poor person is the nigger of the world. But, in truth, Earth itself has become the nigger of the world" (47).

Briefly stated, ecofeminism critiques the practices and beliefs of patriarchal societies (particularly Western industrial society) that have historically identified women and nature as fundamentally "other" and therefore subject to manipulation, control, and exploitation. Gaard explains that ecofeminism's "theoretical base is a sense of self most commonly expressed by women and various other nondominant groups—a self that is interconnected with all life" (1). Ecofeminism identifies that patriarchal and oppressive groups often retain a sense of self that is separate, not interconnected. Many feminist scholars have argued that men are more likely to identify the self as separate from Nature while women are more likely to see the self as connected (see Chodorow 1978 and Gilligan 1982). Ecocomposition embraces not only a greater sense of a connected self, but also a view of connectedness which incorporates all systems, all organisms, all environments. That is, ecocomposition identifies a holistic understanding of systems which affect and are affected by the production of discourse. The notion of a connected self, then, grows not from an inherently Natural self, but from a self connected with and constructed by surrounding environments.

Conceptions of self, according to ecofeminists, are central to defining two ethical systems. According to Gaard, "the separate self often operates on the basis of an ethic of rights or justice, while the interconnected self makes moral decisions on the basis of an ethic of responsibilities or care" (2). This split between connected and separate self, again, are often identified as a male/female division: men focus on rights; women focus on responsibility. According to Gaard the "failure to recognize connections can lead to violence, and a disconnected sense of self is most assuredly at the root of the current ecological crisis" (2). Ecofeminism, then, offers an ethical system based on notions of an interconnected self. Part of such a system requires theorizing oppression on a global scale and political activism toward a variety of issues: "population, global economics, Third World debt, the ideology of development, environmental destruction, world hunger, reproductive choices, homelessness, militarism, and political strategies for creating change globally" (Gaard, 3).

Through the sense of connectedness, ecofeminism has provided important discussions of how environmental oppression are linked with oppression of women which underscore why environmental concerns are feminist issues as well as why feminist issues should be addressed environmentally. One of the most important of these issues is the historical conceptualization of women in Western culture. According to Gaard, "Western intellectual tradition has resulted in devaluing whatever is associated with women, emotions, animals, nature, and the body, while simultaneously elevating in value those things associated with men, reason, humans, culture, and the mind" (5). She goes on to explain that "one task of ecofeminists has been to expose these dualisms and the ways in which feminizing nature and naturalizing or animalizing women has served as justification for the domination of women, animals, and the earth" (5). One of the interesting things about this agenda is that it parallels much of the work that has occurred in composition studies.

In the 1991 landmark essay, "The Feminization of Composition," Susan Miller posits that much of composition's "past, its continuing actual experience, and its usually overlooked but important symbolic associations result from a defining, specifically from a gendered, cultural call to identity" (492). Miller contends, that is, that much of composition's own identity grows from a notion of gendered identity. She explains that

> For many feminist theorists, it is well understood that no matter what range of individual, biological, intellectual, social, economic, class, or other qualities people of the female sex may exhibit, this and other female identities (e.g. "wife," "whore," "girl") participate in similar cultural calls to "womanhood." This "hood" effectively cloaks differences to assure that females (and males) are socially identified by imaginary relations to their actual situations. (492)

She continues to note that "many feminists also point out that within this process, the identity of the female person was specifically differentiated as 'woman' to supplement, complement, oppose, and extend male identity" (493). Using this understanding of cultural labels of gender, Miller argues that this

view of lower-status female identity—including both its critique of dominance and submission and its view of historical requirements imposed for
the sake of survival and tradition—is embodied by composition studies. A
similar cultural call acting on composition has, that is, created the field's un-
entitled "place" in its surroundings and has limited both its old and new
self-definitions. (493)

Miller first supports her argument by providing a historical context for
the development of composition. She notes that composition is "largely
the province of women" and that composition teaching is usually "characterized as *elementary* teaching that is a *service* tied to *pedagogy* rather than
theory" (493). Yet, despite this dominance of the field by women, Miller
notes that in composition research men publish a larger percentage of the
work that appears in *College English* and other publication areas (493–94).
Miller's primary argument is that composition mirrors the cultural oppression of women and "that composition teaching, and composition research,
are not something that 'regular' (meaning powerful, entitled, male-coded,
theoretical) faculty do" (494).

With this argument, Miller, like Cooper, initiates ecocomposition
thinking. Miller's article, in fact, asks that we think ecologically about
composition in two important ways. First, her argument that composition
has been "feminized" in its cultural labeling mirrors the ecofeminist argument that the very hegemonies which lend to the oppression of women
also lead to the oppression of nature. Miller's recognition of the feminization of composition follows the theoretical maneuvers of ecofeminist theory which seeks political action to resist not simply oppression of women
and nature, but oppressions of all kinds. This is crucial to ecocomposition
because it locates ecofeminism and composition as kindred spirits.

Second, what is also important about Miller's position to ecocomposition is that her recognition of the feminization of composition calls into
question larger questions of identity for composition in the university.
Identity is a crucial issue for ecocomposition. As we have noted, we understand that identity comes not only from the social, cultural, ideological
constructs which composition notes as the source for identity, but from
the place where an individual evolves. That is, identity grows from surrounding environments. We know ourselves only in relation to others.
Miller's identification of composition's evolutionary identity is critical to
the ecocomposition agenda as it asks us to question the very relationships
of our discipline to its surrounding environments. Miller's recognition of
composition's role in the university is an ecological understanding. She
forces us to look at composition as a field which is affected by and affects

surrounding environments. She also asks that we question the identity defining not only of composition, but of all organisms based on historical, cultural, and gendered labels.

For ecofeminists, there are at least eight direct links between feminist concerns and environmental concerns, as Karen Warren explains it. First, like Miller's historical assessment of composition's relations to gendered labeling, one of the key facets of ecofeminism is the recognition that such links are historical. However, for many ecofeminists such historical relations are also presented as causal. That is, many ecofeminists argue as Salleh Ariel Kay has that "the current global environmental crisis is a predictable outcome of patriarchal culture" (quoted in Warren 1996, xi). Ecocomposition concurs with this position as it agrees that the very power structures and ideologies which incur oppression over women and other groups are responsible for the oppressive actions taken over nature. Ecocomposition, likewise, contends that colonial activity which spread many of these oppressive ideologies is also responsible for the spread of environmental and natural conquest. Ecocomposition, then, pairs ecocolonialism with ecofeminism in its quest to undermine dominant paradigms that portray "Nature" as an exploitive resource for human use and consumption. The enslavement of Nature is as much a despicable activity as is the enslavement and oppression of women, nonwhites, and, to a less pernicious degree, compositionists.

The second link is a conceptual link. Ecofeminists such as Karen Warren in her essay "Feminism and Ecology: Making Connections," Val Plumwood in her article "Nature, Self, and Gender: Feminism, Environmental Philosophy and the Critique of Rationalism," which appeared in the special issue "Ecological Feminism" of *Hypatia,* has argued that the causal and historical links between the oppression of nature and women are "located in conceptual structures of domination and in the way women and nature have been conceptualized, particularly in western intellectual traditions" (Warren 1996, xi). This conceptualization, of course, parallels Miller's account of the conceptualization of women and composition. In her introduction to *Ecological Feminist Philosophies,* Warren identifies that ecofeminists have noted four primary conceptual links between nature and women. The first of these four has been labeled "value dualism" and identifies pairs as oppositional rather than as complementary, such as the often noted pairs human-and-nature, reason-and-emotion, mind-and-body, nature-and-culture, and of course, man-and-woman. According to ecofeminists the emotion, body, nature, woman side of these pairs are conceived as inferior where as reason, human, culture, and man are given greater value. The second conceptual link of the four that Warren identifies is based on the value dualism and locates such thinking in larger (patriarchal) conceptual frameworks. That is, value dualisms are seen in "all social 'isms of domination,' e.g. racism, classism, heterosexism, sexism, as well as 'naturism,' or the unjustified domination of nonhuman nature" (Warren, 1996, xii). The third conceptual link of the four grows from an understanding that there is a conceptual "basis in sex-gender differences particularly in differentiated personality formation or consciousness. . . . The claim is that female bodily experiences (e.g., of reproduction and childbearing), not female biology per se, situate women differently with respect to nature than men. This difference is revealed in a different consciousness in women than men" (Warren 1996, xii). The fourth conceptual link of the four returns to the historical connections between women and nature and "locates the conceptual link between feminism and the environment in metaphors and models of mechanistic science which began during the Enlightenment and pre-Enlightenment period" (Warren 1996, xiii). This conceptual link identifies that before the seventeenth century, nature was depicted and understood to be likened to a kind nurturing mother, but following the scientific revolution, nature was constructed as machine, nonliving. In both instances, however, nature retained its pairing with "woman." As discourse specialists, ecocompositionists are perhaps best qualified to analyze and explore exactly how language has been used in the manipulation and domination of both women and the biosphere.

The third overall link many ecofeminists identify is one of empirical and experiential evidence. This link often identifies the health risks imposed on women by pesticides, toxins, radiation, and other pollutants which often have adverse effects on women and children in disproportionate levels. Empirical data often reflects such claims and are critical to this type of ecofeminist research. Often, this data is distorted, obfuscated, and even ignored by groups and corporations whose economic or political interests might be threatened by such knowledge. Ecocompositionists working in environmental rhetoric, such as those found in Herndl and Brown's *Green Culture* (more on this later) are able to clarify and elucidate such knowledge, and their work can often result in various forms of public intellectualism.

The fourth link is an epistemological one. Because of the inquiry into other links between feminism and environment, feminist environmental epistemologies have had to evolve. One of the most important discussions of the evolving epistemologies in ecofeminism comes from Plumwood's *Hypatia's* article. In "Nature, Self, and Gender: Feminism, Environmental Philosophy, and the Critique of Rationalism," Plumwood contends that environmental philosophy has failed to "engage properly with the rationalist tradition, which has been inimical to both women and nature" (3). She continues to argue that "current brands of environmental philosophy, both those based in ethics and those based in deep ecology, suffer from this problem, that neither has an adequate historical analysis, and that both continue to rely implicitly on rationalist-inspired accounts of the self that have been a large part of the problem" (3). In her critique she claims that

> The problem is not just one of restriction *in* ethics but also a restriction *to* ethics. Most mainstream environmental philosophers continue to view environmental philosophy as mainly concerned with ethics. For example, instrumentalism is generally viewed by mainstream environmental philosophers as a problem in ethics, and its solution is seen as setting up some sort of theory of intrinsic value. This neglects a key aspect of the overall problem that is concerned with the definition of the human self as separate from nature, the connection between this and the instrumental view of nature, and broader *political* aspects of the critique of instrumentalism. (10)

It is specifically these political aspects of the nature-human binary which Warren contends a feminist environmental epistemology would address.

The fifth link is symbolic. Like ecocritics who look toward the textual representations of nature as affecting human attitude toward nature, many ecofeminists inquire as to the ways in which literary, religious, theological, artistic, and other textual representations of nature and women affect the treatment of women and nature. Similarly, many ecofeminist turn to more linguistic approaches to the representation of women and nature as inferior. Ecocompositionists study the ways that discursive practices are negotiated in power and politics, and they highlight the fact that discursive representations of women and nature often result in oppressive actions toward both.

The sixth link addresses the ethical connections between women and nature. According to Warren "the claim is that the interconnections among the conceptualizations and treatment of women, animals, and (the rest of) nonhuman nature require a feminist ethical analysis and response. Minimally, the goal of feminist environmental ethics is to develop theories and practices concerning human and the natural environment which are not male-based and which provide a guide to action in the prefeminist present" (1996, xv).

The seventh link identifies that discussions which make the connections between feminist issues and issues of environment have produced a broad range of theoretical approaches to exploring such links. For instance, as Warren notes, this range of theoretical positions is most evident in the area of environmental ethics. There is not, according to Warren, "one ecofeminist philosophical ethic" (xvi). While Warren's analysis does not mention ecocomposition specifically, issues of feminism, issues of the environment, and the production and use of discourse in such matters are all deeply imbricated and are central to the project of ecocomposition.

The eighth link requires political action and is often called *praxis;* that is, the eighth connection is a call to action. This call to action grows from the fact that many ecofeminist concerns evolved from a variety political concerns ranging from issues of health, technology, animal rights, anti-nuclear movements, and so on. One of the most significant aspects of eco-composition is its dedication to various aspects of public intellectualism. A primary motivation for many scholars working in ecocomposition is the degree to which it negotiates issues that have real global implications.

Without question, ecocomposition finds many important connections with ecofeminism, and ecocomposition aligns itself with many of the agendas of ecofeminism. For instance, ecocomposition exposes oppressive ideologies much in the same way feminist theory or race theory do. Eco-composition identifies the historical and patriarchal ideologies which have cast nature and environment as subservient to (hu)man needs. There is lit-tle question that these oppressive positions grow from the same ideologies which cast women and other groups in the same light. Ecocomposition seeks to identify the ways in which discursive constructions lead to and can resist such traditions. That is, ecocomposition asks as to how the produc-tion of written discourse contributes to the positioning of nature in con-temporary culture.

Similarly, ecocomposition recognizes that the very production of dis-course which contributed to such historical constructions are also respon-sible for the conceptualizations which code nature as both feminine (i.e. mother nature) and as resource. Ecocomposition seeks to overturn concep-tualizations which demarcate natural environments as entities which may be mastered and ruled. In turn, ecocomposition resists oppression of all living organisms—human or other—and their environments. Ecocompo-sition specifically asks as to the role written discourse has played and con-tinues to play in the "othering" of natural environments.

While ecofeminism turns to empirical data to identify correlations between environmental and feminist issues, ecocomposition inquires as to the ways in which such evidence is presented and used politically. That is, ecocomposition, while aligned with contemporary ecology and other envi-ronmentalist and natural sciences and is supportive of empirical research, turns to questions of postmodern science (see Sandra Harding 1986 and Donna Haraway 1991) to question how such evidence is amassed and then used. Certainly, ecocomposition sees personal experiences crucial to the ecomposition agenda as does ecofeminism. For ecocompositionists, separating personal experiences in any environment are inseparable from the experience of writing. All writing grows from personal encounters with environment. Environment is, after all, context. While ecocomposition situates itself in the social constructionist camp and identifies that writing is a social process and that all knowledges are created socially (including nature), it also identifies that personal awareness and experience of environments—natural and human-made—are necessary in order to produce successful writing.

Drawing from this intensely connected notion of environment and discourse, ecocomposition embraces ecofeminism's (and ecocriticism's) inquiry into the symbolic and textual representations of nature. Though, as mentioned earlier when discussing ecocriticism, at its core, ecocomposition is more concerned with how environment affects the production of discourse rather than the interpretation of how nature is represented in text. Of course, separating the two agendas are nearly impossible, and studying textual representation of nature is critical to the understanding of how textual production occurs.

While ecocomposition owes a great deal to ecofeminism, it also hopes to move beyond its focus on gender to address all of the many ways in which human beings dominate and exploit the biosphere. Although ecofeminists generally focus on *androcentrism,* or male-centeredness, as the basis for the human exploitation of nature, scholars including Warwick Fox, H. Lewis Ulman, and Roderick Nash suggest that male-centeredness is just one part of a larger problem: *anthropocentrism,* or human-centeredness. Fox suggests that while men have been implicated in the history of ecological destruction far more than women, we must extend this examination to account for the actions and attitudes of all white, wealthy Westerners, who have been far more implicated in global environmental destruction than precapitalists, blacks, and non-Westerners. He argues that "anthropocentrism has served as the most fundamental kind of legitimation employed by *whatever* powerful class of social actors one wishes to focus on" (22). The work of ecofeminists, environmental ethicists, cultural critics, and ecocompositionists, through their examinations of the discursive practices that enable anthropocentric thought, might begin to extend our sense of community to include nonhuman nature. What makes human beings unique, in Joseph Kockelmans' words, is our ability to apprehend the world and our place in it as a "totality of relations" (9). However, without discourse, we stand mute: we can experience our world, but without the ability to articulate its messages we cannot act to preserve this totality of relations.

Likewise, ecocomposition is a call to action. The ethics of ecocomposition must evolve into epistemological positions, must move to sites of action that resist those currently in place. Ecocomposition should be seen not as a way one kind of (nature) writing gets done, but as how all writing is produced. Ecocomposition must not be solely a classroom or solely an academic endeavor. Ecocomposition must develop many theories from many perspectives. It must account for a wider range of positions and it must question the patriarchal domination of nature. One of the areas in which such activism has grown and contributed to ecocomposition is through the study of environmental rhetoric.

Environmental Rhetoric

Separating composition from rhetoric is an impossible task—and rightly so. Like composition studies in general, ecocomposition is deeply enmeshed with rhetorical studies. If we are to claim that ecocomposition is interested in the exploration of the production of written discourse as it relates to environment and place and that ecocomposition grows from both ecological and environmental concerns, then part of ecocomposition's definition falls within the boundaries of rhetorical studies as well. Ecocomposition is concerned with rhetorical analysis of environmental/political issues, the affects of language on those issues, and the ways in which ongoing debates or conversations affect the ways in which writers write. Political debates, activism, and participation all rely on a rhetoric of the environment which is of critical importance to ecocomposition.

Two texts, in particular, have aided in the development of ecocomposition's understanding of ecological/environmental rhetoric: M. Jimmie Killingsworth and Jacqueline S. Palmer's *Ecospeak: Rhetoric and Environmental Politics in America* (1992) and Carl G. Herndl and Stuart C. Brown's *Green Culture: Environmental Rhetoric in Contemporary America* (1996). Both of these texts takes on the project of critical analysis of the discourse through which environmental issues have been conveyed by addressing contemporary issues in environmental politics through the lens of rhetoric.

M. Jimmie Killingsworth and Jacqueline S. Palmer begin *Ecospeak: Rhetoric and Environmental Politics in America* by noting that

> Since the middle of the last century, human beings have become increasingly aware of the earth's vulnerability. Examining the results of humankind's technological powers has opened a new vein of consciousness, the knowledge that large-scale human action may place the further existence of nature—

including human activity itself—in jeopardy. With this new awareness have come calls for "extensive social readjustment" as well as the initial rumblings of "violent social upheaval"—the historical conditions that, according to Charles Morris and Edward Corbett, give rise to an invigorated practice and study of rhetoric. (1)

In the pages that follow, Killingsworth and Palmer offer a primarily analytical discussion of "the patterns of rhetoric typically used written discourse on environmental politics" (1). They base their rhetorical readings in alignment with Charles Bazerman's claim that "histories of rhetoric have too exclusively focused on *rhetorica docens,* the theory and pedagogy of rhetoric, while ignoring 'actual living practice'" (1).

Foregrounding *Ecospeak,* is the notion that "the environmental dilemma is a problem generated by the way people think and act in cultural units" (2–3). Killingsworth and Palmer contend that "since human thought and conduct are rarely, if ever, unmediated by language and other kinds of signs, it is understandable—possibly inevitable—that rhetorical scholars enter the environmental discussion through the gate of humanism" (3). Ecocomposition embraces and aligns itself with the notion that human thought and conduct are (most likely) always mediated through language, and it identifies such mediation as the source for the construction of nature, place, environment, and the conduct and behavior directed at those locations.

Killingsworth and Palmer explain that "ecologically conscious humanism tends to portray current environmental problems as a crisis of Western Liberalism" (3).

> Briefly put, the dilemma is this: How can the standard of living attained through technological progress in the developed nations be maintained (and extended to developing and undeveloped nations) if the ecological consequences of development are prohibitive? . . . This confrontation has led to some serious questions about what constitutes the good life in a progressive culture and about how to reconcile new scientific knowledge with material progress based on information generated by a scientific culture of the recent past. (3)

Thus, for Killingsworth and Palmer, the environmental dilemma is one of "Western society's inability to resolve ethical problems" and extended "because the ethical problem issues from a crucial epistemological problem—humankind's 'alienation from nature'" (4). Killingsworth and Palmer contend that in this issue of environmental ethics and epistemology we must consider not only the "rules for community disputes," but we also "must additionally understand how the disputants construct their views of the natural or nonhuman worlds" (4). Often, then, "one group will view nature as a warehouse of resources for human use, while an opposing group will view human beings as an untidy disturbance of natural history, a glitch in the earth's otherwise efficient ecosystem" (4).

Ultimately, as much as issues of ethics are central to environmental debates, Killingsworth and Palmer note that the environmental dilemma is a problem of discourse:

> Various proposals to resolve the crisis are put forth by different social groups with different sources and kinds of information, groups with divergent goals, methods, values, and epistemologies. All groups have a particular perspective and use a specialized language developed specifically to describe and stimulate the practices characteristics of their particular outlook on the world. (7)

The goal of this specialized language, or internal rhetoric, according to Killingsworth and Palmer, via Kenneth Burke, is identification:

> Following Burke, we can think of this problem in grammatical and narrative terms. Any action may be stated by an active voice sentence, the kernel of a group's identifying story: *I (or we) do this.* In the case of intactable problems, the subject position of one group ("we") cannot be filled with members of another group ("you" or "they"). Rhetorical appeals propose enlargements of the *we* category or mergers of the two categories, with the ultimate goal being the identification of the "global" public with the "local" discourse community. (7–8)

Ultimately, then environmental dilemmas reflect "the current historical stage of ecological consciousness; the environmentalist ethos only now in the process of formation so that the conflicts we feel are a part of the inevitable process by which political stance evolves" (8). Killingsworth and Palmer's investigation rightly explores the degree to which the environment, and how we think about, discuss, and act upon it is largely dependent upon discourse.

Similarly, Carl G. Herndl and Stuart C. Brown's collection *Green Culture: Environmental Rhetoric in Contemporary America* takes as its foregrounding notion that "the environment is not a thing you could go out and find in the world. Rather, it is a concept and an associated set of cultural values that we have constructed through the way we use language" (3). This social constructionist view of environment contends that "there is no objective environment in the phenomenal world, no environment separate from the words we use to represent it. We can define the environment and how it is affected by our actions only through the language we have developed to talk about these issues" (3). This assessment of the discursive construction of environment lies at the cornerstone to ecocomposition's understanding of environment. For Herndl and Brown, access to understanding this construction, in its various incarnations is provided by rhetorical analysis of particular conversations and debates. Rhetoricians, according to Herndl and Brown, "study the ways people use language to construct knowledge and to do things in the world" (vii). *Green Culture* contains eleven essays which take as their task to do just that: study the ways in which language is used to construct various environmental debates. The editors note in their introduction that

> The environment about which we all argue and make policy is the process of the discourse about nature established in powerful scientific disciplines such as biology and ecology, in government agencies such as the Environmental Protection Agency and its regulations, and in nonfiction essays and books such as Rachel Carson's *Silent Spring* or Paul Erlich's *The Population Bomb*. . . . The values and beliefs we hold about the environment are established through the discourses of a bewildering variety of genres, institutions, and media. (3–4)

They go on to say that "the language of these various discourses determines what exists, what is good, and what is possible" (4). For Herndl and Brown, the "field of environmental rhetoric is immense and remarkably varied, so varied in fact that we think it connects almost every part of our social and intellectual life, crossing the boundaries between various academic disciplines and social institutions" (4). However,

> for rhetoricians who study the way we use language to construct our world and to conduct our lives, this wide range of environmental discourse is both interesting and problematic. The variety of very different contexts in which we talk about the environment suggests that there is not one environmental discourse but many, a polyphony that makes it difficult to understand and resolve environmental disputes. (4)

For ecocomposition, this notion of multiple contexts of environmental rhetoric is crucial as ecocomposition focuses on ways in which those varied contexts affect the very production of discourse. Ecocomposition turns heavily to rhetorical critique of the multiple environmental discourses for the understanding of how environmental rhetoric is composed and how writers contribute to those discourses which Herndl and Brown identify as the location where environmental issues are addressed.

However, ecocomposition also extends Herndl and Brown's understanding of the multiple contexts for environmental discourse to claim that all discourse is already environmental and that while rhetorical critique of particular utterances are of benefit to understanding how to "understand and resolve environmental disputes," ecocomposition must look beyond environment as merely a thing about which we have disputes and about which discourse participates and creates, but as the very thing that the production of all discourse is reliant upon and contributes to. The editors of and contributors to *Green Culture* provide some insightful rhetorical readings of nature writing, policy and law, and the history of environmental rhetoric, all of which participate in larger conversations about the interaction between discourse and environment.

By addressing such concerns, these investigations into environmental rhetoric intersect with several important aspects of ecocomposition. First and foremost, they explore connections between discourse and environment, and as such, they clearly serve an important task within the concentric environments of ecocomposition, composition, academia, and the world itself. They suggest that discourse is integral to mapping, shaping, and constructing the world, and in turn, they see language as a powerful tool in eliciting social, political, and environmental action. Environmental rhetoric is a powerful pedagogical approach that serves a number of purposes: it raises student awareness of the ways people use language to construct knowledge and accomplish things in the world; it allows students to see that language is a powerful tool that influences us and, in turn, can be used to influence others; it enables them to better recognize ways in which different discourse communities structure the content, form, and rhetorical appeals of their language to better communicate with their intended audiences; and last but certainly not least, environmental rhetoric informs students of global issues that are among the most important in the world today. As we and our students struggle to make sense of the deluge of information surrounding environmental issues, analyzing environmental rhetoric provides a means to organize and clarify much of this information.

Composition and Nature

Just as ecocriticism, ecofeminism, cultural studies, and environmental rhetoric have been crucial in the evolution of ecocomposition, so too has composition theory played a primary role in the development of ecocomposition. There are many facets of writing theory which lend to our understanding of the relationships between discourse and environment and the construction of environment, place, nature, location, and so on. One of the key notions which composition theory provides ecocomposition is the idea that all discursive construction creates, maintains, defines, and reinscribes the ways in which we perceive nature. As Stan Tag writes, "When we study the relationships between language and landscape, text and terrain, or words and woods, we are not studying two separate things (as if we lived in some dualistic universe), but interdependencies, particular manifestations (even processes) of the thing we call life, each interconnected to the other, and both wholly dependent upon such basic natural elements for their survival as sunlight, water, and air" (1998). For ecocompositionists this interconnectedness is crucial. Composition's theoretical moves post-process contribute greatly to our understanding of these relationships.

When cognitivists in composition began examining the processes by which individual students compose written text, this inquiry was undertaken with little regard for the sites in which writing takes place. That is, cognitive process understanding of writing was reluctant to consider the affect environment had on those very process. As composition moved away from cognitive models of writing, compositionists began to consider the implications external forces might have on writers, and in turn what affect those very writers had on those same external forces: gender, culture, race, class, ideology, and the likes. As Hawisher, et. al., explain, "During the period of 1983–1985, composition studies absorbed the changes brought about by the new emphasis upon process and began to chart the course it would follow post-process, looking beyond the individual writer toward the larger systems of which the writer was a part" (65). In essence, composition began an ecological approach to understanding discourse. Instead of looking at individual organisms (writers), compositionists began to examine the relationships writers have with other writers, with culture, with race, with gender. Similar to ecology, composition began to see writers as contributing members of larger systems. However, even in composition's move post-process there was little actual recognition of environment beyond theoretical understandings of ideology and other constructed critical categories. The actual places where writing develops and occurs were not considered in the post-process move. Ecocomposition continues the post-process move to understand relationships between writers and larger systems by taking into consideration the role of environment, of place, of nature, of location in those larger systems by examining the relationships between discourse and place.

Donald Davidson's theory of triangulation contends that in order for an individual to come to know the world, or objects in the world, that individual must participate in a discursive moment with another individual and the world. That is to say, we come to know the world through our contact with others and that very world. As Davidson explains it:

> Each of two people finds certain behavior of the other salient, and each finds the observed behavior of the other to be correlated with events and objects he finds salient in the world. This much can take place without developed thought, but it is the necessary basis for thought and language learning. For until the triangle is completed connecting two creatures and each creature with common objects in the world there can be no answer to the question whether a creature, in discriminating between stimuli, is discriminating between stimuli at the sensory surfaces or somewhere further out, or further in. It takes two to triangulate. For each of us there are three sorts of knowledge corresponding to the three apices of the triangle: knowledge of our own minds, knowledge of other minds, and knowledge of the shared world. Contrary to traditional empiricism, the first of those is the least important, for if we have it we must have the others, so the idea that knowledge could take it as foundation is absurd. (quoted in "Production," 65–66)

More succinctly, as Anis Bawarshi explains it: "We come to know and understand objects in the world and each other only when our interpretations match others' interpretations" (73).

Elsewhere Sid has argued about the power relationships in discursive triangulation, and so we want to note here that triangulation is not innocent, that hegemony and oppression begin at the moment of triangulation.[3] But here, what we want to note is the social, discursive manner through which we come to know the world. For human beings, we cannot know the world, or as we have noted, we cannot know nature, except through language. And just as the triangulative moment proves to be the moment at which power takes hold—the moment at which one participant in the triangulative event may lead the other to a particular understanding—so too is that the moment at which humans construct, map, and elicit power over nature. But before we come to this point, let us be more explicit as to the workings of triangulation.

David R. Russell, in his article "Vygotsky, Dewey, and Externalism: Beyond the Student/Disciple Dichotomy" offers a simplified example of a triangulative moment:

[A] seven-month-old child who has not yet learned her first words reaches in the direction of a spherical object and babbles. Her parent, seeing this, puts the object in her hands and says, "Ball! You want to play with the *ball*?" Sooner or later—usually sooner—the child learns that adults may play with her using spherical objects and that certain sounds ("ball") and certain activities accompany human interactions with such objects. She learns through observing others' actions and her own that making the sound "ball" in certain situations often produces certain effects in others. Triangulation has been achieved. And learning.

The child will eventually learn many words and effects for many kinds of spherical objects and many kinds of activities to go with them. But the crucial point here is that a linguistic system or conceptual scheme or community norm or discursive convention did not mediate between the child's mind and the object. Another human being, the parent, mediated between ("triangulated with" might be a better phrase) the child and the object. (181)

Granted, Russell's example fails to recognize the gender roles of child and mother, the power structures already established between parent and child, and a host of other problems. Yet, as Russell and Davidson see it, the child does not come to know "Ball" until it has experienced "Ball" in a discursive moment with another and come to some consensus as to what ball is, what ball does, and who has right over ball. The same is true for the relationship between humans and nature. Simply put, we do not know nature, or know something as natural, until we have reached such consensus with another, or at the very least with ourselves.

Those of us who work with ecotheories and have more "natural" concerns must also question encounters of nature through individual experience and solitary discovery. That is, what of the moment when an individual comes to know part of the world in a solitary moment? Much like ecofeminism, then, ecocomposition recognizes the importance of individual experience, but identifies those moments as triangulative as well. Thomas Kent, drawing from Davidson's triangualtion theory, contends that discourse is a non-codifiable system, that no two moments of discursive interaction are like any other. Hence, he argues that humans develop prior theories which take into account all past discursive experiences and provide us with theories for each next interaction. If this is the case, then humans who discover on their own, do so not through some divine, romantic inspiration, but by accessing past discursive experiences. For extreme example, we can easily contend that the Tarzan myth doesn't really happen, that small human children are not raised by forest animals with no human contact. Most any human that would find itself alone in the wilderness has already developed some prior theories, because that individual has had previous discursive experiences with others. As a result, that individual is never truly "alone" because he or she has with them the memories of prior discursive contacts. These discursive memories enable the individual to triangulate with an other, even though that other or others is not actually present. For instance, on a more realistic note, if one were to spend a good deal of time exploring a particular patch of nature, one might make many discoveries, observe many wonders. However, each of those discoveries, each way that individual names what is found is dependent upon what that individual has known before. A new tree is known to be a plant, a tree, a shrub, or something else, and that knowledge carries with it a host of other associations bound up in prior theory.

But what of that new tree and the individuals coming to know it alone on a hillside? No one else is present with whom to triangulate. No one else reaches consensus with the individual about what that object in the world is. Triangulation still occurs. Only now the triangulative moment occurs between self, object, and self. We triangulate with ourselves (this is a legal activity, even in the Bible Belt). We examine an object, ponder it, and come to some consensus with ourselves as to what that object is. And though we may not reach any definitive conclusion, we may still wonder about it; we have enough of a conversation to agree with ourselves that it is a plant or a fish or an insect or a fungus or a rock. Furthermore, we use our prior discursive contacts with others to generalize and (to some degree) classify that object. For instance, we notice that the object is green and leafy, and, based upon our previous knowledge of objects of this sort— knowledge gained through contact with others—we surmise that this object is some sort of tree. At all moments, then, nature as we know it is a discursive formation.

If we are to understand triangulation, then, as the communicative moment at which we come to know the world, then the instance of triangulation also dictates how we come to know nature. That is to say, nature is defined discursively. For us, nature does not exist outside of the discursive formations we assign to it. That is not to say that "Natural" systems do not exist, but only that they occupy the positions into which we place them discursively. For instance, our choice of naming something as "habitat" or "endangered" or "natural" or "man-made" is a discursive—and often a politically dangerous—maneuver which identifies status within nature. For example, in Florida, the manatee is a protected species. The manatee population is counted, watched, and cared for when individuals are injured. In the winter months, when water temperatures drop and some manatees, which are extremely temperature sensitive, die, the local news stations carry stories of the deaths as tragic events. Manatees, like porpoises and whales, are ranked high in the "cute-factor," so they are constructed as things to "save" and things which we attribute human qualities such as intelligence and emotion. A mere three hours by plane from Florida, in Central America, however, manatees are constructed differently. They are food. They are hunted. In fact, a mere forty years ago, manatees were constructed as food in Florida as well. A manatee does not occupy the same position in nature in Florida as it does elsewhere in the world solely because of the discursive construction applied culturally and historically. The same holds true for a variety of food species around the world: cows in India, whales in Japan, dogs in Korea. To be overly simple about it, nature does not exist until we name it as nature. Should our prior theories identify that a particular object in the world, an oak hammock for instance, is part of the natural world, then it is given a particular status different from the status we might provide a car wash or an elevator. That is, nature, à la Burke, steps out of language. According to Neil Evernden, "The act of naming may itself be a part of the process of establishing a sense of place" (101).

In other words, this act of naming constitutes a place as having particular boundaries, particular functions, and particular identities. For example, in Hawaii, as in many areas that have indigenous populations, the naming (or more accurately, the re-naming) of particular locations is of great concern, particularly if those areas once held sacred status. Ancient *heiaus,* or religious sites, have been and are being destroyed to make way for shopping malls, condominiums, and resorts. Naturally, these sites are also often renamed, giving them Westernized names. Many Hawaiians protest both the physical and the discursive changes taking place in these areas, arguing that both the places themselves and their names reflect Hawaii's social and cultural history. For these protesters, the place is inseparable from its name; erasing the name erases history, erases identity, erases place. Of course, this example reflects the renaming and redefinition of a place from one cultural perspective to another. Much of the naming of locations worldwide reflects the urbanization and development of places that were once relatively free from human exploitation. We've already mentioned that the Florida state park system has adopted the motto The Real Florida. Of course, the motto is intended to evoke a sense of old, wild Florida. The very construction (literally) which sets aside a particular site as identifiably "real," as opposed to the constructed streets and strip malls found adjacent to many parks, creates a discursive relationship between what can and cannot be part of nature. If an object falls within the boundaries of the fenced area, it is not only natural, but it is also protected. If it falls outside the boundaries of the park, it may not be natural, or it may be

developed as it is unprotected. The implications of naming something as a
"real" part of Florida, are critical to our understandings of nature, and in
turn, to our understandings of discourse. The same holds true for any sort
of mapped environment: classrooms, universities, cities, states. As Edward
Abbey has pointed out, "the itch for naming things is almost as bad as the
itch for possessing things" (256).

> Through naming comes knowing; we grasp an object, mentally, by giving it
> a name—hension, prehension, apprehension. And thus through language
> create a whole world, corresponding to the other world out there. Or we
> trust that it corresponds. Or perhaps, like a German poet, we cease to care,
> becoming more concerned with the naming than with the things named;
> the former becomes more real than the later. And so in the end the world is
> lost again. No, the world remains—those unique, particular, incorrigibly
> individual junipers and sand stone monoliths—and it is we who are lost.
> Again. Round and round through the endless labyrinth of thought—the
> maze. (257)

Language of Nature

In his journal entry of November 2, 1833, Ralph Waldo Emerson writes:

> Nature is a language, and every new fact that we learn is a new word; but
> rightly seen, taken all together, it is not merely a language, but the language
> put together into a most significant and universal book. I wish to learn the
> language, not that I may learn a new set of nouns and verbs, but that I may
> read the great book which is written in that tongue. (149)

Ecocomposition identifies that nature is a text, not only a text that we
may learn and read, but one we may write, or more precisely, one we may
write in, on, with, or about. As Emerson would later write on November
10, 1836, "Language clothes Nature" (150). Yet, contemporary culture has
relinquished its understanding of nature as language. Certainly, postmodernity identifies nature as text, but only in the sense that nature is something to which critical readings might be directed. That is, the notion of
nature as text, still denies nature its own voice. At the University of Florida, for instance, a flyer for the organization Students for the Ethical
Treatment of Animals (SETA) asks volunteers and activists to attend a
rally to "speak for those that have no voice." The assumption is that nature has no voice.

Earlier we noted ecocomposition's agreement with Killingsworth and Palmer's claim that "human thought and conduct are rarely, if ever, unmediated by language and other kinds of signs" (3). Ecocomposition also contends that all organisms mediate their conduct and thought through systems of signs (though we note the difficulty in assessing thought or measuring such claim in all organisms in any reliable manner, since doing so would require an anthropomorphic filter). Ecocomposition listens to the language and signs of nature as text which can and should be interpreted. That is, ecocomposition identifies languages and signs other than human languages as critical to the notion of ecological communication and seeks to give voice to nature in its own right rather than simply as a subject about which humans assign voice. We agree with Christopher Manes "Nature is silent in our culture (and in literate societies in general) in the same sense that the status of being a speaking subject is jealously guarded as an exclusively human perogative. The language we speak today, the idiom of Renaissance and Enlightenment humanism, veils the processes of nature with its own cultural obsessions, directionalities, and motifs that have no analogues in the natural world" (15). Similarly Kenneth Burke writes in this now famous example that

> We may begin by noting the fact that all living organisms interpret many of the signs about them. A trout, having snatched at a hook but having had the good luck to escape with a rip in his jaw, may even show by his wiliness thereafter that he can revise his critical appraisals. His experience has led him to form a new judgment, which we should verbalize as a nicer distinction between food and bait. A different kind of bait may outwit him, if it lacks the appearances by which he happens to distinguish "jaw-ripping food." And perhaps he passes up many a morsel of genuine food simply because it happens to have the characters which he, as a result of his informing experience, has learned to take as a sign of bait. I do not mean to imply that the sullen fish has thought all of this out. I simply mean that in his altered response, for a greater or lesser period following the hook episode, he manifests a change in behavior that goes with a new meaning, he has a more or less educated way of reading the signs. It does not matter how conscious or unconscious one chooses to imagine this critical step—we need only note here the outward manifestation of a revised judgment. . . . Our great advantage over the sophisticated trout would seem to be that we can greatly expand the scope of this critical process. (5)

As William Howarth notes, "Unlike hunters or writers, nature makes direct statements, without implication or analysis" (71). Anyone who has spent a good deal of time in nonhuman environments knows this to be accurate. The message of a rattlesnake's rattle or a cotton mouth's hiss does not leave much room for interpretation; the message is clear and succinct. Misinterpretation has great material ramifications. The message spotted and striped skunks send are easily read. The message of a palmetto spear is to the point. The coloration of an IO moth sends a message of warning, as does the posture or grunt of an alligator. Nature not only speaks, it composes. A bear or fox writes its territory, marking it with scent and markings—markings which may be "read" by other animals. When tandem jumping porpoises perform their acrobatics, humans read an aesthetic of precision and playfulness, other porpoises read what is being written: a warning.

As Howarth notes, "Ecology leads us to recognize that life speaks, communing through encoded streams of information that have direction and purpose" (77). If we are to arrive at a fuller, more ecological understanding of discourse, we might begin by interrogating some of our most basic presumptions regarding language itself. We might begin to recognize, for example, that our very *conceptions* of discourse are human-centered. Nearly all of Western thought tends to construe language as that power which humans possess and other species do not. We presume that the distinguishing feature of humanity is our ability to think and communicate our thoughts. Since the first Greek *physiologoi,* or natural philosophers, who named, classified, and ordered the natural world, language has been claimed as the exclusive and distinguishing property of humankind; we have always presumed privileged access to *logos.* Our conceptions of discourse, rest upon a narrowly anthropocentric system of belief. We disagree with Edward O. Wilson's premise that only *Homo sapiens* are capable of constructing language that is productive and capable of conveying information. Wilson suggests that the language instinct consists of "precise mimicry, compulsive loquacity, near-automatic mastery of syntax, and the swift acquisition of a large vocabulary" (145). How convenient that this definition corresponds precisely to human language. How interesting too that Wilson himself has published landmark research regarding the chemical communicative methods of ants through pheromones and the physical communicative methods of bees.

At least one contemporary linguist has called this system of belief into question. In his book *Language and Human Nature,* Harvey Sarles argues that the assumption that language is a purely human property provides little more than an ideological justification for the human domination of nature. In fact, Sarles suggests that this narrow conception of language makes it impossible for us to comprehend the nature of our own discourse itself.

Our inability to properly contextualize language and see it as something beyond verbal utterances (or marks on paper or a computer screen) limits our conceptions of how language is created and used. Language is not a strictly human endeavor; only by recognizing that other life forms communicate in and through their environments will we be able to more adequately theorize our own roles as language users. Sarles asserts that "to define language as uniquely human also tends to define the nature of animal communication so as to preclude the notion that it is comparable to human language" (86). He suggests that language must be understood in much broader terms than it currently is, if we are to fully recognize the importance of discourse in the construction of knowledge.

> Each ongoing species has a truth, a logic, a science, knowledge about the world in which it lives. To take man outside of nature, to aggrandize the human mind, is to simplify other species and, I am convinced, to oversimplify ourselves, to constrict our thinking and observation about ourselves into narrow, ancient visions of human nature, constructed for other problems in other times. (20)

Sarles' perspective opens us toward an understanding of the subtle relationships between language, ecologies, and the construction of knowledge. Until we recognize that we hold no monopoly on language, and that our language has an ecological dimension, we will be unable to fully conceptualize how knowledge is formed. David Abram suggests that "as long as humankind continues to use language strictly for our own ends, we will continue to find ourselves estranged from our actions" (97). Adopting a more ecological conception of language, then, can be seen as an integral component in our quest for a more holistic understanding of how we and other life forms communicate and generate knowledge.

Ecocomposition and Public Intellectualism

In order to achieve many of the goals outlined above, ecocomposition must become a site not only for public intellectualism, but also for what we will call in chapter 5 "activist intellectuals." Unless ecocomposition moves beyond the borders of the academy and its politically safe environments of classrooms and textbooks, ecocomposition stands to fare no better than any one of a dozen theoretical and practical trends which never find voice beyond the institutional walls of academe. Ecocomposition places environment as a central concern of theory, pedagogy, and *praxis*. In the process, it raises the preservation of natural environments to its rightful place as one of the most significant political concerns in composition studies today. Like ecofeminism, ecocomposition asserts that "To gender, race, and class, we should now add nature, and in the process create a contextual web of social, political, and natural concerns" (McAndrews, 377).

Ecocomposition stresses two principles that are imperative to successful public intellectualism: our interdependence on others and our interconnectedness to a larger biosphere. Public intellectualism involves (or should involve) these two principles as well. Successful public intellectualism necessitates a sense that we are all dependent on shared resources. To this end, we might work toward language that incorporates the interests of diverse members of our local and global ecosystems—including other intellectuals, other peoples, and even other organisms. The public or activist intellectual, then, can be seen as one who thinks inclusively, works toward common goals, and speaks and writes in ways and through avenues that takes as many "others" into consideration as possible. The activist intellectual also recognizes that we are all connected to a series of larger spheres, including the academic community, but also including the cities, states, and countries in which we locate ourselves and the larger biosphere of planet Earth. This recognition allows us to situate our concerns and voices as part of these larger spheres, and as such, it might help us to see our disciplinary and institutional boundaries as less important than our goals. This type of thinking helps to erase territorial battles over particular academic areas and moves toward creating more inclusive conversations. Such thinking also allows us to recognize the diverse groups and individuals that we are connected to.

The environment, which has been at the forefront of many current public debates, has been defined through the words we use to represent it: it is the product of a variety of discourses and their intersections in the public sphere. Many ecocompositionists have begun to realize that studies in environmental discourse, for instance, are both important to us as human beings and valuable to us as rhetoricians. Most of us come to ecocomposition through our love for natural places and our love of discourse. Activist intellectualism allows us to combine these two loves and extend our boundaries beyond the classroom. By working through the classroom, through our scholarship, and through public work, we can speak and act effectively on environmental issues both locally and globally. Ideas and efforts move through movements such as ecocomposition. The efforts of activist intellectuals should be comprised of more than any one type of work. As members of academia, we have the social authority to make significant changes in our communities and, perhaps over time, in our culture. If we begin to refigure our roles not as public intellectuals (as the term has come to be defined), but as activist intellectuals and understand that activism is open to all of us, we can begin to have a significant affect on the ways humans interact with and in the living biosphere. We take up these issues in detail in chapter 5.

Ecocomposition as Pedagogy

While our work as public intellectuals often takes us outside of the academy, we return to what has too often been painted as the home place of composition: the classroom. While composition theory has developed in sophisticated and meaningful directions over the past forty years, it has, unlike theory in some other disciplines, nearly always been done in an effort to make our work in the classroom more effective. Composition is unique in that it revolves not around a particular body of knowledge, but around the common goal of helping students to become better writers. The project of knowledge-making in composition is deeply implicated in how teaching practices are formed and argued for. As Joseph Harris writes in *A Teaching Subject*, "composition is the only part of English studies which is commonly defined not in relation to a subject *outside* of the academy (to literature, for example, or to culture or language) but by its position *within* the curriculum—by its close involvement with the gatekeeping first-year course in writing" (xi). Similarly, work in ecocomposition often returns to the kind of day-to-day practices that go on in college classrooms and departments under the rubric of reading and writing.

Ecocomposition pedagogy must address several primary goals. First, it should address the production of written discourse. That is, ecocomposition pedagogy must take as its primary pedagogical agenda the teaching of writing. In all ecocomposition courses, issues which directly assist students in becoming better producers of writing in a variety of writing environments must be the central focus. Ecocomposition should also encourage students to recognize their experiences in all environments as effecting and being effected by their writing. Ecocomposition requires hands-on experiences in a range of environments. In turn, students should be taught to be critical of how those places are mapped, defined, regulated, and managed through discourse in order that they may identify for themselves how discourse affects and is affected by places they experience and that the find connection with. Nature and environment must be lived in, experienced to see how the very discourses in which we live react to and with those environments. According to Emerson:

So lies all the life I have lived as my dictionary from which to extract the word which I want to dress the new perception of this moment. This is the way to learn Grammar. God never meant that we should learn Language by Colleges or Books. That only can we say which we have lived. (152)

Ecocomposition pedagogy must encourage political activism, public writing, and service learning. That is, student writing should be directed beyond the limited scope of classroom assignments to address larger, public audiences. Students should be encouraged to use their writing to affect change, to bring about awareness, to promote their own political agendas. Students should be encouraged to become active. Passive learning neither provides students with writing scenarios in which they encounter real audiences, nor does it promote ecological awareness and participation. We explore in greater detail the possibilities of ecocomposition pedagogies in chapter 6.

Moving from Composition to Ecology

Many of us working in ecocomposition have moved to this area of study through careers in composition studies. That is to say, ecocompositionists are, for the most part, compositionists who have brought their concerns for environmental protection and ecological literacy to composition classrooms and composition research. We have yet to identify the ecologist whose interest in writing has led him or her to ecocomposition. In essence, ecocomposition is born from composition, grounded in composition, and occupied by ecocompositionists. Our knowledge of ecology, of environment often comes from personal study and research, not formal training. The conversations we have with other ecocompositionists generally take on the same framework as those conversations we have with any compositionists. Our understandings of ecological sciences are often cursory, picked up in reading along the way. Yet, composition, in its claim toward interdisciplinarity is, like most of the humanities, wary of the sciences. We still acknowledge the boundaries of our own and other disciplines and critique scientific methodologies and agendas (as well we should). However, in recent conversations with ecologists, biologists, and environmental scientists, we have also begun to question how those of us in the humanities see the sciences differently from how scientists see the sciences. For ecocomposition, this means asking ourselves explicitly as to what we really understand about ecology as a science. We cannot suppose that a cursory reading of a few landmark texts in ecology/environmental science could yield any particularly important new insight for composition; we must learn much about these other disciplines if they are to inform our own. That is to say, if ecocompositionists are to draw upon ecology as more than just a metaphor or a conceptual tool for thinking about writing, then we must study ecology and its methodologies.

While our own investigations into ecology and environmental sciences have been far from exhaustive, we have found much of what we've read, studied, discussed, and analyzed to be rewarding and useful. Most ecocompositionists agree that theories and studies in ecology and related fields offer new insights and ways of thinking about the relationships between individuals, discourse, and environments. For that reason, in chapter 3, we examine a few of the insights we have gleaned from ecological and environmental sciences in an effort to initiate new and potentially innovative conversations in composition and ecocomposition alike.

CHAPTER 3

Ecology and Composition

꧁※꧂

In every perception of nature there is actually present the whole of society.
—Theodor Adorno, *Aesthetic Theory*

Ecology is a science strongly connected to a history of verbal expression.
—William Howarth, "Some Principles of Ecocriticism"

In the language of ecology, the biosphere is a conversational domain. The languages in which its discourses unfold are written in the genotypes and expressed in the phenotypes of myriad organisms.
—Daniel R. White,
Postmodern Ecology: Communication, Evolution, and Play

One of the overriding misconceptions regarding ecocomposition—much as is the misconception about ecology—is that ecocomposition is an environmentalist application whose sole purpose is the protection and preservation of natural environments and their inhabitants. It is not. Or, more accurately, it is *not only*. Certainly, many (in fact, most) ecocompositionists align themselves with particular environmental issues, and many are drawn to ecocomposition primarily because it foregrounds the importance of *places*—both natural and human-made. In addition, ecocomposition is often seen as an opportunity to teach texts that address these issues. That is to say, most frequently, ecocomposition is assumed to mean the study of nature/environmental writing. However, the fundamentals of ecocomposition that we have laid out in chapters 1 and 2 identify that ecocomposition allows for theoretical exploration, practical engagement, pedagogical treatment, and also environmental/ conservationist/preservationist activism. In other words, ecocomposition is not a study of nature writing but a study of writing and ecology and the ecology of writing. In order for ec-ocompositionists to put forth worthwhile theories and pedagogies about what ecocomposition might be or become, we feel it is both useful and profitable to understand a bit more about the origins of ecological think-ing. We don't propose to speak for any community of scientists or scholars who might define themselves as ecologists. We are compositionists. We, along with most ecocompositionists, work within departments of English and envision ourselves as discourse specialists. But like many ecocomposi-tionists, we have found more than a few intersections, similarities, and af-finities between the study of discourse and the study of ecology. For this reason, we put forth the following study to help to make these connections a bit more apparent.

Ecology in the Western Tradition

Since ecocomposition takes its name from a melding of ecology and com-position, it seems most appropriate to begin by exploring ecology, since for compositionists, it is likely to be the less-familiar root (sorry) of ecocom-position. *Ecology* is a term that is often misused, often uttered in place of *environmentalism,* and often tagged as a leftover liberal movement from the 1960s. Ecology certainly falls within the rubric of the environmental sciences, as do geology, climatology, biology, genetics, evolution, and a host of other sciences, but it is not a science grounded in environmentalist politics as the contemporary mis-use of the word often suggests. Ecology's etymological roots lay in the Greek *oikos,* meaning "home" or "house." This is the same Greek root from which "economy" also is derived. Ecol-ogy also derives from *logos,* literally from *discourse.* The term was first used (in modern times) by Henry Thoreau in 1858, but it was not clearly defined until German zoologist Ernst Haeckel offered in 1869 that "oecology" is the study of all relationships between an organism and its organic and in-organic environments. In other words, ecology seeks to explain how living creatures work with and against one another for resources and space. Haeckel, a follower of Darwin's theory of evolution, believed that the term would help to better define the struggle for existence which Darwin ad-dresses in *The Origin of Species.* According to Haeckel:

> By ecology, we mean the body of knowledge concerning the economy of na-ture—the investigation of the total relations of the animal both to its or-ganic and inorganic environment; including above all, its friendly and in-imical relation with those animals and plants with which it comes directly

or indirectly into contact—in a word, ecology is the study of all the complex interrelationships referred to by Darwin as the conditions of the struggle for existence. (quoted in Ricklefs, 1)

For the most part, contemporary ecology still maintains this basic definition. Many ecologists have noted that if we are to understand ecology in
terms of Haeckel's definition then, as ecologist Charles J. Krebs notes,
"there is very little that is *not* ecology" (3). Compositionists must also recognize that there is very little in our current conceptions of post-process
composition studies that is not ecological. That is, when composition
studies began to explore the relationships between writers and exterior
forces that effect and are effected by writers, composition studies began an
ecological look at discourse, and very little of what we now do in composition studies is not ecological.

What is interesting about Haekel's definition is that it smacks of the
links to economic notions of cost, benefit, and waste. Haeckel was most
likely also influenced by his contemporary Karl Marx, since Haeckel's definition of ecology appeared just two years after the first publication of *Das
Kapital*. It is almost certain that he had read Marx's work or at least was
very familiar with its basic premise. Regardless of its origin, the metaphor
of economics has been pervasive throughout ecology's short history. In
fact, much of the scholarship and teaching tools of the discipline still rely
heavily on this metaphor. Donald Worster's landmark book *Nature's Economy: A History of Ecological Ideas,* as well as one of the most often used
textbooks for college-level introduction to ecology classes, *The Economy of
Nature* by Robert E. Ricklefs reflect the omnipresence of this metaphor.

The term *ecology* did not come to general use until the late 1800s when
a few German and American scientists adopted the term. Professional
journals and organizations for ecologists did not appear until the early
1900s. Yet, as ecologist Donald Worster notes, the study of ecology "is
much older than the name" (378). The history of ecology is as old as the
history of human communication. In fact, ecology and writing evolved together, for the same purposes. Early forms of logogram writing, such as the
famed cave paintings of Altamira or those of the Anasazi (and others found
throughout the world), Sanskrit writings, and hieroglyphs depict the mi-
gratory routes and seasons of food species, identify records of agricultural
endeavors and planting seasons, and detail the knowledge retained about
hunting, fishing, and food gathering.[1] Early Babylonian and Egyptian
writers wrote of locust swarms and other natural events. That is, writing
began as a means by which to record, count, calculate, codify, and taxoni-
mize human relationships with nature. In fact, it has been argued that the

history of writing is in a sense the history of civilization, for without the ability to communicate complex information, no society would have been able to develop to any great extent. As Jay David Bolter suggests in *Writing Space,*

> Writing is a technology for collective memory, for preserving and passing on human experience. The art of writing may not be as immediately practical as techniques of agriculture or textile manufacture , but it obviously enhances the human capacity for social organization. . . . Writing is and has always been a sophisticated technology: skill is required to learn to read and write, and penetrating intelligence is needed to invent or improve some aspect of the technology of literacy. (33)

In other words, the rise of sophisticated methods of communication has been seen as a precursor to "sophisticated" societies. Bolter also gives specific examples of this correlation between the emergence of complex language use and complex societies, suggesting that "the earliest economies flourished in Mesopotamia and in Egypt, where picture writing was gradually replaced by phonetic systems, in which written symbols were associated directly and consistently with sounds in the language" (37). However, it also must be noted, as does Jared Diamond in his Pulitzer Prize winning book *Guns, Germs, and Steel: The Fates of Human Societies,* that writing was "restricted geographically: until the expansions of Islam and of colonial Europeans, it was absent from Australia, Pacific Islands, subequatorial Africa, and the whole New World, except for a small part of Mesoamerica" (215). Diamond is also correct to note that while many have attributed writing as a characteristic of "civilization," "some peoples (notably the Incas) managed to administer empires without writing" (215). However, if, as we have just suggested, in the writing that did evolve first, those first complex transmissions of information dealt with matters of ecology—again, the relationships between living things and their environments—it is easy to see that writing and ecology have always been deeply imbricated. Writing (or discourse, more accurately) was (and is) the first tool of ecology. If ecology is the science by which relationships to environment are measured for the purpose of management—that is, for the purpose of economizing environments—then writing is the vehicle through which ecology is employed. However, it must also be noted that just as ecology sought to manage nature for the benefit of human purveyors, so to did writing become a colonial tool. In other words, in many ways, ecology developed to give humans a better control over nature; it emerged as a system for colonization and oppression. Similarly writing became a tool for the spread of knowledge, and as composition studies has argued for many years now, writing (and language in general) is perhaps the most powerful tool used by oppressive regimes. As Diamond puts it, "writing marched together with weapons, microbes, and centralized political organization as a modern agent of conquest" (215–16). To say that ecology and writing evolved hand in hand, one must acknowledge also that writing and ecology conquered and colonized hand in hand as well.

While writing and the study of ecology afforded the rough tools for (the advancement of) certain advanced civilizations, these tools were best employed by the precursors of Western thinking—the ancient Greeks. Like those early Egyptian and Babylonian writings, Aristotle also wrote of swarms of locusts and their relationship with certain environments. In *Historia Animalium,* he explored the relationships between prey species such as field mice and predators such as foxes and ferrets. In fact, much of the ancient Greek way of life was based on ecological harmony, a recurring philosophy among early cultures and religions: Native American, Norse, Hindu, Buddhist, Incan, and Polynesian, to name but a few.[2]

Discourse—particularly written discourse—has always mapped out human relationships with nature and environment. For instance as Christopher Manes writes, "Many primal groups have no word for wilderness and do not make a clear distinction between the wild and domesticated life, since the tension between nature and culture never becomes acute enough to raise the problem" (18). In the 1998 Academy Award winning film *As Good as It Gets,* Greg Kinnear's character—a wounded artist—is awestruck by the beauty of Helen Hunt's character and in a moment of inspiration insists that she "is the reason cavemen painted on walls." Though a romantic line, it seems more likely that cavemen painted on walls, sandstone, papyrus, or any other material in order to begin to understand and catalog the relationship of humans to the natural world. That is, written discourse evolved as a means by which to record, study, and manage human relationships with nature. To essentialize, writing and ecology developed for the very same reasons. Composition is as much an ecological project as ecology is a discursive, cultural project. Composition is as much a project grounded in economics as is ecology. Both are primarily concerned with consumption and production.

In fact, if we are to extend Manes' understanding of culture's need to not identify separation between nature and culture, we can begin to understand that a particular environment leads to particular ways of life which, in turn, leads to the development of culture. Early civilizations, for instance, were dependent upon relationships with local environments, hence culture grew *from* and *in* environment. In many cases, the knowledge of local environments was regarded as sacred and religious leaders were trusted with keeping that information sacred. Medicine men and shamans often held the duty of maintaining knowledge of growing seasons and animal migrations.[3] In fact, many early religions grew to define relationships between natural and human entities in order to help preserve and understand this very knowledge. Egyptians had to contend with the regular flooding of the Nile, an act they attributed to the gods providing for them. The water of the Nile sustained human life. For the Egyptians, the floods were supernatural events created by gods. However, the Greeks first attempted to identify a rational understanding of humans' relationships with the world. In his introduction to *Minding Nature: The Philosophers of Ecology*, David MacAuley writes that "Western philosophy begins as a meditation on nature in an attempt to discern the order of things and to speculate on its meaning, direction, and purpose" (1). For instance, the flooding of the Nile was seen by Greek geographers not as an act of gods, but as an effect of seasonal rains in Africa. In Western thinking, the Greeks are the starting point with which explanations of "nature" become issues of rational science rather than supernatural acts of gods.

Of course, at the core of Greek philosophy was the concept that the earth stood at the center of the universe and that the four elements—earth, air, fire, and water—composed that world. Many Greek thinkers sought to define models of the universe that incorporated geology, chemistry, biology, and even evolution. The first Greek *physiologoi*, or natural philosophers, reflected "not simply on the human *psyche* but directed themselves foremost toward the yawning heavens, the turnings and reversals of fire, the rhythm and play of water, and the outcroppings of rock and earth—more broadly, the four elements" (1). Thales of Miletus, for instance, maintained that water was the element of which all living things were made and that earth itself floated in water. His thinking, of course, mirrors earlier religious beliefs which identify Poseidon as the god of water; his effects were felt on dry earth because water was seen as responsible for earthquakes. Thales simply concluded that the waves of the supporting water would crash into the Earth and cause earthquakes. Similarly, his student Anaximander identified wind as the central element and related lightning and clouds to wind. He argued that all matter could be traced to the basic substance of the cosmos: air (Bowler, 34–40). Many other Greek philosophical and rational explanations followed: Hippocrates' *Airs, Waters, Places* stands as prime example.

Perhaps one of the most important discussions of *Airs, Waters, Places* is that highlighted by David Arnold in *The Problem of Nature: Environment, Cultural and European Expansion*. Arnold notes the division of *Airs, Waters,*

Places into two distinct sections and suggests that though the two portions may have been penned by two separate authors, they have been of critical importance to environmentalists and ecologists alike. The first section of Hippocrates' text addresses medicine and physiology in relation to environment. Summarizing this first part, Arnold writes that "its avowed purpose is to help a doctor understand the causes of the diseases he is likely to encounter in moving to a new locality—a district, for instance, that is exposed to northerly winds, or that draws its water from stagnant ponds and marshes rather than from free-flowing springs, a place where the seasons, soils and vegetation may be very different from those with which he is familiar" (15). Arnold continues to explain that Hippocrates saw this as vital knowledge for a doctor, since environment had a direct affect on human health.

> It is assumed that all human beings are basically much alike: what makes them different are the environmental forces, the airs, waters and places, and hence the diseases, to which they are exposed. In Greek thought, the human body was often perceived as a microcosm of nature, and thus agitated nature, like a cold North wind or the change from season to season, was bound to have a corresponding effect upon human physiology. (15)

The second part of *Airs, Waters, Places* compared the environments of Asia and Europe in order to discern between racial and cultural difference. According to Hippocrates:

> Asia differs very much from Europe in the nature of everything that grows there, vegetable or human. Everything grows much bigger and finer in Asia, and the nature of the land is tamer, while the character of the inhabitants is milder and less passionate. The reason for this is the equable blending of the climate, for it lies in the midst of the sunrise facing the dawn. It is thus removed from extremes of heat and cold. Luxuriance and ease of cultivation are to be found most often when there are no violent extremes but when a temperate climate prevails. (quoted in Arnold, 15–6)

Throughout his discussion, Hippocrates notes the importance of culture in creating social atmospheres, but environment remains central to his position: "the constitutions in the habits of the people follow the nature of the land where they live" (quoted in Arnold, 16). Of course one can easily identify the problems of inherent racism in Hippocrates' position as he often makes claims such as

The small variations of climate to which the Asiatics are subject, extremes both of heat and cold being avoided, account for their mental flabbiness and cowardice as well. They are less warlike than Europeans and tamer of spirit, for they are not subject to those physical changes and the mental stimulation which sharpen tempers and induce recklessness and hotheadedness. Instead they live under varying conditions. Where there are always changes, men's minds are roused so that they cannot stagnant. (quoted in Arnold, 16)

Thankfully, the attribution of such totalizing racial stereotypes have been set aside (for the most part), but Hippocrates' claims can be seen as the first notion of ecological perspective which first formulates, according to Arnold via Clarence J. Glacken "that human minds, bodies, even whole societies, were shaped by their geographic location, their climate and topography" (17). This is critical for those of us in composition studies, particularly for those of us who have adopted cultural-studies methodologies. Hippocrates ultimately informs contemporary cultural studies that we must also consider environment in examining the construction of culture. We are not suggesting that we return to Hippocrates' inherently racist definitions of culture as defined by environment, nor are we suggesting a reductionist view that culture is nothing more than an environmentally defined phenomenon. What we are suggesting is that we must look at place and environment as central to the evolution of culture and, in turn, culture's reciprocal relationship with discourse. Compositionists have embraced notions that account for the degree to which discourse and identity arise as a result of the intellectual and social climates in which they are born. Social constructionist perspectives of composition dominate the field, and through these, we've come to better understandings of how individuals emerge as a result of a number of social forces. Like cultural studies approaches to composition, social constructionist perspectives suggest that social forces shape the style, form, content, and degree of complexity of the discourse that individuals and groups use to communicate. As ecocompositionists, we feel it necessary to more fully account for the degree to which place and environment directly and indirectly affect discourse as well. We'd like to extend the social constructionist notion of the group or the community as the fundamental category in shaping knowledge and discourse to include the larger framework in which these groups operate—their material environments. That a world exists beyond the confines of specific cultures is an idea that seems almost rudimentary, and yet it is an idea that has been for the most part overlooked by social constructionist theory. As Christian argues in "Ecocomposition and the Greening of

Identity," "The world that we perceive and inhabit, which we often forget in favor of the human culture it supports, is always a part of who we are and how we express ourselves" (86). We argue here that all subjectivity, all awareness, all knowledge and discourse presupposes our inherence in an enveloping world we call "Earth," and the differences and nuances (as well as the similarities) between groups and individuals are greatly affected by specific physical ecologies.

As we have explained in chapter 2, cultural studies, particularly composition's version of it, has only recently begun to consider place in the construction of culture and identity and has done so in very limited ways. Drawing from Hippocrates, ecocomposition contends that environment is of great importance in the construction of identity, and that in fact, culture grows out of environment. The place where a culture, where a discourse evolves, has tremendous effect on the evolution of that discourse or culture. In fact, in many ways environment has a more powerful force upon culture than does discourse, than does ideology, simply because ideologies and discourses grow from environments. Of course, to return to the chicken and egg argument we began regarding culture and discourse and environment in chapter 2, because environment is also a social construct, it becomes problematic to suggest that environment precedes culture, discourse, or ideology. In fact, while we are playing this game of chase our theoretical tails, it is important to note that the concepts of wilderness, nature, place, and environment are strictly discursive constructions which are products of Western thinking, for example, human/nature dualism, emerging directly from Aristotle, Hippocrates, and other Greek thinkers. Nonetheless, Hippocrates' initial claim to the influence of environment on culture stands as important in the history of the development of ecocomposition, though his claim is by no means ecocomposition's sole foundation.

The philosophy which stands most likely as a foundation for contemporary Western thinking regarding nature was forwarded by Xenophon, a student of Socrates, in his text *Memorabilia*. Xenophon argues that the Earth is a physical system which is responsible for sustaining human life and that Earth's environment, animals, and plants are designed to support human life. Xenephon's philosophy, however, identifies that such designs must have been created by a *human-like* being who *transcends* and *controls* the natural world. This philosophy stands as precursor to Christian teleological approaches to the understanding of nature. Such an approach to nature also became a primary characteristic of Stoic philosophy which saw nature as generated to maintain order and purpose in the world. Combining Stoic approaches to nature with teleological approaches, Panaetius and Posidonius later contended that while humankind exists in nature and are rightfully exploitive of nature, nature is also intended to teach humans (Bowler, 44). The classic example is the invention of a ship's rudder which was designed based upon observations of fish fins and tails or the observation of birds' wings in the development of airplane wings.

Aristotle extended the teleological understanding of nature. Aristotle, of course, is still regarded as one of the first natural historians and his explanations of natural systems are still critical to contemporary thinking. Aristotle's thinking about nature grew in his split from Plato who resisted scientific investigation of nature. For Plato, philosophy was the avenue by which the mind could be freed from the material world. According to Plato, truth and knowledge were achieved by identifying connections between ideas, not from attention, observation, or thinking about the material world. Physical objects were inferior representations of ideas. Bowler points out that only in *Timaeus* does Plato discuss creation and offer his views on the relationship between thinking, ideas, knowledge, philosophy and the material world (46–7). In this text, Plato initiates the principal of "plenitude" in which he argues that nature becomes the expression of the mind. In other words, Plato first recognized that human thinking constructs nature. Aristotle breaks with Plato because he wished to develop a philosophy more attentive to the material world. At the core of Aristotle's philosophy were natural history and natural processes.

Aristotle categorized the Earth into different regions or *klima,* which he devised according to the amount of the sun's heat that reached particular regions. According to Aristotle, life can only be maintained in certain *klima:* those receiving too much sun such as equatorial regions, he argued, were too hot to support life. Conversely, those areas that received little sun, could not sustain life because they were too cold. In *Meterologica,* Aristotle pays close attention to various natural systems and processes and addresses topics such as earthquakes, volcanoes, wind, caves, rivers, rain, flooding, and agriculture. And yet, Aristotle maintained a teleological position because he believed that the processes of nature could not create the world or the universe, but that natural processes could be studied. As Bowler notes, Aristotle specifically contends that there is a difference between possibility and actuality: a seed is not a plant, but the potential for a plant (50). For Aristotle, nature maintained organization over matter. As Bowler explains it:

> Aristotle understood all constructive change in terms of four causes: the formal (the structure being created), the material (the matter upon which form is being imposed), the efficient (the actual force working on the matter) and the final cause (the purpose for which the new structure is created).

Teleology (explanation in terms of purpose) was essential to his philosophy, but he did not neglect natural limitations that must be imposed on how the purpose is filled. (51)

Seen from this perspective, Western thinking can be seen as an essentially ecological pursuit. As we've suggested, the early Greek philosophers attempted to discern the order of nature and to speculate the place of humans in it. They saw no distinction between the study of nature and the study of knowledge, and the concepts of mind and nature were virtually indistinguishable. Thought could not be distinguished from science. According to David MacAuley, "Aristotle and his student Theophrastus can be seen as early forerunners of ecological thought—the Greek fathers of animal and plant ecology, respectively" (1). Through the speculative acts of definition and classification, they attempted to make sense of the world and describe the relations between plants, animals, and humans. To a remarkable extent, they integrated scientific observation of nature with philosophic inquiry and analysis. They envisioned a "kosmos" which ecologized thought by recognizing that it could not be separated from its place in the material world.

Western thought since that time has been, in many respects, a dissolution of the connections between man in nature. Theories of nature gradually gave way to theories of the mind. The Scientific Revolution's intellectual will to dominance, and especially dominance of the natural world, can be seen as the culmination of dualistic thinking which separates the mind and nature. Most notably, this can be found in the empiricism of Sir Francis Bacon and the rationalism of René Descartes—two Renaissance thinkers who characterized the natural world in terms of its usefulness for and distinction from humans. In fact, many of the contemporary discursive positions regarding nature and man have their origins in the work of sixteenth- and seventeenth-century Enlightenment thinkers.

For Bacon, to understand nature means to disturb it and alter it *(natura vexata)* by means of technology and/or ingenuity so that it will give up its treasures. In the advancement of his kind of thinking, Bacon asserted that humans are separated from nature and that the earth exists for the use of humans. Bacon argued that man maybe regarded as the center of the world,

In so much that if man were taken away from the world, the rest would seem to be all astray, without aim or purpose . . . and leading to nothing. For the whole of the world works together in the service of man; and there is nothing from which he does not derive use and fruit . . . in so much that all things seem to be going about man's business and not their own. (quoted in Marshall, 184)

For Descartes, the certainty of the self can only occur after separation with nature. After a clear separation of the human psyche from the environment which surrounds it, precise measurement and carefully reasoned scientism is possible. The history of Western thought after Descartes continues to echo his intent to "make ourselves masters and possessors of nature" (quoted in Rifkin, 32). Cartesian thinking envisions a clear split between discourse—which for Descartes is a human creation that arises from within the mind—and nature—that which is separate from the human mind. In other words, for Descartes, discourse and environment exist as polar opposites, the former being a means through which to control the latter. In keeping with this, John Locke asserted that the "negotiation of nature is the way to happiness" (quoted in Strauss, 315). Clearly, the Scientific Revolution's intellectual project remains central to much of our culture's thinking. This begins to explain the relative lack of ecological and environmental perspectives in contemporary thinking.

Ecologies and the Production of Discourse

As we mentioned earlier, the term *ecology* did not come into use until the late 1800s, and it was not until the early 1900s that journals and organizations evolved to support the work being published by scientists who called themselves ecologists. In the last seventy years or so, ecology has established itself as a significant branch of the environmental sciences, and the science itself has grown, changed, and redefined itself several times. Yet, for the most part, ecology maintains itself as a managerial science. That is, many ecologists see ecology as having the potential to solve many of the problems created by the massive increase of human population on the earth and the strain that population has put on Earth's resources. According to Robert E. Ricklefs, "management of biotic resources in a way that sustains a reasonable quality of human life depends on the wise application of ecological principals to solve or prevent environmental problems and to inform our economic, political, and social thought and practice" (2).

Basically outlined, ecology observes organisms and habitats at various levels to determine the relationships between various ecological systems by which, according to Ricklefs, "we mean any organism or assemblage of organisms, including their surroundings, united by some regular interdependence"

(2). Ecological systems, then, operate much in the way that discourse communities do, or more accurately, discourse communities can be likened to ecological systems. Like the attempt many in composition have made to identify ways in which writers participate in and are conscious of various discourse communities and/or discursive systems, ecologists seeks to identify a variety of levels of organization in ecological systems. For ecologists like Ricklefs, "the organism is the most fundamental unit of ecology" (2). There is no smaller unit of life (such as a cell or molecule) which has a separate life (Ricklefs does note the single-celled organism as being an exception, but refers to these as organisms, not cells). The organism, its structure and its function are "determined by a set of genetic instructions inherited from its parents and by the influence of many factors in its environment" (3). Ecosystems, then, are groups of organisms which function together in a particular environment (physical and chemical) and exchange energy within the system in order to metabolize, grow, and reproduce. In order to do so, organisms often manipulate their very environment and the amount of energy available to them and other organisms found in the ecosystem. And, according to Ricklefs, "all ecosystems are linked together in a single biosphere that includes all the environments and organisms on the surface of the earth" (3).

The simple link, then, to connect ecological thinking with composition is to simply revisit composition studies in an ecological light. Writers (or users of discourse) are the fundamental unit operating within these systems. Writers manipulate environments by writing and finding ways for their writing to fit within systems. Writers' structures and functions are determined, not by genetics per se, but by knowledge and ideology which functions much as DNA does; ideology and culture map (if not control) our thinking and actions much like a genetic code. A writer can no more easily escape the ideology of the discourse community in which he or she operates than an animal or plant can escape its own particular ecosphere. Certainly both can adapt, often quite successfully, but that adaptation nearly always comes with great difficulty. Writers are, as ecocompositionists and social constructionists argue, also influenced by environment and, in turn, influence that same environment. That is, while ideology and culture map our thinking, our environments shape the application of that thinking. Much like Darwin's finches and tortoises and his theory of evolution, writers enter into particular environments with a certain ideological (genetic) code and then contend with their environments as best those codes allow. As the environment shifts, writers readjust their characteristics to match the environment. Much like genetic evolution, writers display certain characteristics in their writing that are determined by the environments in which they write. That is, a genotype offers a set of genetic instructions which are manifested through a phenotype, or an expression of those instructions, so, too, does an ideology offer a set of ideological instructions which are manifested in the use of discourse. Of course, genetic evolution allows for organisms to develop over time new genotypes or characteristics that have been developed or altered because of environmental conditions and expressed phenotypically, so, too, do users of discourse react to environments and maintain the potential to alter ideologies. Discourse communities become ecological systems in which writers interact with and react to one another and their environments. And, all writers are linked in a single discourse-sphere where we identify that no discourse community exists free of other discourse communities. Such an ecological view of the compositionist's world also sheds light on the theories of paralogic rhetoric and discursive triangulation we mentioned in chapter 2 and will continue to explain in chapter 6. Just as our prior theories provide the information needed to formulate passing theories—the very theories we use in individual communicative moments—so, too, do attributes and characteristics alter with environments in order to become genetic systems. In other words, the adaptations and adjustments we make as speakers, writers, listeners, and readers are not genetic adjustments; they do, however, mirror the manner in which organisms evolve to better survive within changing ecosystems. And though such an ecological approach does provide an enlightening metaphor by which to see composition studies, doing so only over-simplifies the larger relationships between the two cultural projects of ecology and composition studies. That is, ecology and composition are concerned with theorizing and cataloging how complex systems interact— be they organic, inorganic, or discursive—and viewing composition through an ecological lens even though ecology does not provide full access to composition's ecological methodologies.

Ecology begins with the basic premise that organisms live within *environments,* often referred to as *habitats.* Habitats can be classified by their physical characteristics: terrestrial or aquatic, for instance, and may then be subdivided by more specifics: forest and deciduous forest or marine and pelagic. Habitats may be classified to extremely specific details, constituting microhabitats. Such classifications are identified for the sole purpose of study, for paying attention to specific boundaries of a particular habitat. In Florida, for instance, ecologists may look to the Everglades as a specific habitat they wish to study, or they may identify a more specific area such as the mouth of Graveyard Creek. Within these areas, ecologists examine the organisms which inhabit these places and the effects the environment may have on those organisms. At Graveyard Creek, for instance, ecologists may examine the numbers of Everglades Palm trees, a species found only within the borders of Everglades National Park. Ecologists may measure the amount of light which is filtered out from taller trees to discover how much light reaches and affects the Everglades Palm. They may look for evidence of predation. They may compare information with that gathered about the Florida Silver Palm, a species native to south Florida, but with a range extending beyond that of the Everglades Palm. Based on this information, ecologists then establish a range of tolerance, the boundaries of the habitat in which the species can live and survive. Similarly the ecologist has his or her own range of tolerance which we might call a discourse community. The very ecologist who might have defined the range of tolerance of the Everglades Palm is also an organism living within an environment, bound by its constraints, dependent upon its resources. Ecologists must also exist within a particular community in order to be identified as ecologists. That is, ecologists can only exist within a discursive community that allows an ecologist to survive; we might call this environment academic or scientific discourse. The ecologist must negotiate this environment, which is mediated by the ideology of the community and which comes with certain frameworks for questions and inquiries which identify an organism as an ecologist. Certainly our ecologist, like the Everglades Palm, must compete for limited resources. Where the Palm competes for nutrients, water, and sunlight within its community, the ecologist competes for employment, grants, and publication space in journals and presses. The ecologist, in fact, has needs to explain, understand, and question, and she or he is reliant upon a certain set of resources provided by an environment, just as a plant has certain needs. Similarly, the ecologist needs many of the same things a plant needs even before she or he enters the discursive environment that allows him or her to be named an ecologist. All needs must be met in order to survive.

Ecologists discern between abiotic factors—factors of nonliving influence—and biotic factors—relationships with other living creatures which affect the life of the studied organism. Organisms may use one another for food; organisms may compete for food or space. And within these relationships, organisms function within their range of tolerance. Organisms react to and with their environments, often changing, altering their habitats. All habitats, like all discursive habitats, are in a state of continuous flux: seasons vary; wind, precipitation, and temperature induce change; erosion, growth, depletion, all contribute to the change. To return to our comparison, a particular strain of palm may do better or worse than similar strains due to its ability to conserve nutrients or water, just as the ecologist may do better or worse within his or her academic community due to the interest and importance (to the discipline) of his or her particular specialization. Importantly, both the palm and the ecologist can affect the environment in which they exist; the palm's use of water, nutrients, and sunlight may change the area surrounding it, making it more or less difficult for other plants and animals to thrive in its particular ecosystem, and the ecologist may alter his or her particular academic ecosystem (through discourse) to the advantage or disadvantage of other ecologists.

Organisms react to change. Organisms may respond physiologically, altering body temperatures, hibernating, increasing photosynthesis all to maintain some constant within the organism itself, to maintain homeostasis. Organisms may respond behaviorally: sensile organisms (those which are bound to one location such as some plants) may rotate leaves or close petals; motile organisms (those which are mobile) may exhibit more visible behavior changes, such as the recent increase in Tiger Shark attacks on humans in Hawaii, speculated by some biologists to be caused by decreasing populations of open-ocean prey. Physiological changes lend themselves to a critical aspect of an organisms survival: acclimation. According to Ricklefs, "Acclimation is a reversible change in structure that helps maintain homeostasis in response to environmental change" (219). That is, organisms must respond to changes in environment in order to survive; they must acclimate and they must maintain the ability to reverse those changes, to return to environments as they become more hospitable to habitation. Acclimation is the ability to change and change again; acclimation is the act of revision. In Florida, many species of fish have the ability to adjust their tolerance of salinity levels in waters. The redfish, for instance, has the ability to move from salt water to brackish water to freshwater in order to move into warmer waters of Florida's spring-fed rivers in the winter when water temperatures drop.[4] It then has the ability to reverse its changes and return to salt waters. Similarly, writers must acclimate, must adjust to particular writing environments. Organisms must acclimate as their environments change; all writers must revise as their environments, purposes, and audiences shift. As Brewer puts it, "Acclimation is undoubtedly of great importance in allowing organisms to exist permanently in changeable environments. Without this ability many organisms would either die or be forced to migrate during unfavorable seasons" (9). The same holds true for writers in a variety of political and ideological environments. Certainly, we have seen political environments which force acclimation under fear of death. Fortunately, as we will discuss in a moment, such political and ideological environments can be resisted (to some extent) and writers may wield their ability to change such environments.

Ecocomposition identifies that the physiological and behavioral changes which organisms undergo in order to maintain homeostasis mirrors the maneuvering a writer must undertake in order to survive in a writing environment. Few writers ever exist within the metaphorical center of an academic community; most writers are constantly working to gain a greater foothold within the discourse community in which they are writing. While this is easily seen in regard to first-year writing students, who attempt to learn "academic discourse" in order to survive their next four or so years in college, the same holds true for most scholars and teachers of writing, who hope to secure their place (and ensure their survival) in the academic community through publishing scholarly books and articles. That is not to say, however, that ecocomposition strives for homeostasis among writers. To make such a claim would be to argue in favor of hegemony. What ecocomposition recognizes is that homeostasis is necessary to some degree. That is, in order for writing to be effective, its production must occur within the parameters of local environments; it must know its own range of tolerance. Writing must conform to certain preinscribed rubrics in order for it to function. Writing must use a particular (accepted) discourse; in fact, it must use prescribed language and conventions. Particular communities and styles (genres) of discourse provide the environment in which writers operate. Genres function like ecosystems; they define those very ranges of tolerance. For instance, if we were to begin a sentence "thsjhjo ijfdjsd wxv," even though we have used recognizable symbols of the English alphabet, that sentence would not survive here. If we were to acclimate, to some degree, and continue our sentence with "blue thug brat fish," our acclimation would move toward homeostasis in that we now employ a recognizable vocabulary. Yet, that vocabulary must fit within even more stringently defined requirements of the writing environment. We must write sentences that are as near discursive homeostasis as we can produce. It is important to note, however, that while ecosystems often fall within specific parameters and usually have certain codifiable guidelines, complete homeostasis within one of these ecosystems is nearly impossible. For example, while it may be easy to identify the boundaries of a particular ecosystem—the subarctic zone of Mauna Kea, or the Kau Desert (both on the Big Island of Hawaii)—it would be less easy, perhaps impossible, to identify a particular individual or species that is most suited and adapted to it. Similarly, it may be easy to identify composition as an intellectual ecosystem bounded by particular subjects, journals, and conferences, but it is impossible to name individuals or even specializations within composition that are most suited for success within the discipline.

What then of resistance writing, of revolutionary writing, of abnormal discourse? What of *ecriture feminine*, a writing with an intention of resistence of homeostasis, of resistence of hegemony? What ecocomposition identifies is that homeostasis is not hegemony, that the agendas of resistence writing like *ecriture feminine* is one of migration. That is, resistence writing stands to move away from hegemonic environments and move toward homeostasis with a newly defined environment. Many organisms, rather than acclimating through physiological changes, opt for migration—a move to more hospitable environments. Yet, like *ecriture feminine*, those migrating organisms must still operate within their range of tolerance, though that range may be vast. Writers, too, may range far, but they must still operate within the larger environment of writing; *ecriture feminine* is first and foremost *ecriture*—writing—(the noun described by the adjective *feminine*). A Kemp's Ridley sea turtle, for instance, may spend summer months on Florida's Gulf Coast, but when conditions change in late autumn, it migrates to warmer waters. Yet, at all times the Kemp's Ridley must remain within its aquatic, shallow coastal environment. It cannot migrate to Kansas (though it may be forceably transplanted there, but that's another matter concerning the reality of artificial environments and survival in those environments).[5] Similarly, writers must remain within the boundaries of writing to survive as writers. As we have noted, all organisms alter their environments. The mere fact that an organism exists within an environment necessitates that the organism affects and is affected by that environment. Resistance writing is the conscientious altering of writing environments. It is the recognition that certain environments pose a threat or are not suitable for habitation. Resistance writing is the purest form of ecological homeostasis as it is the writing which moves toward total homeostasis—a homeostasis without competition for survival. Yet, all ecosystems, all writing habitats must also be biodiverse. Certainly, resistance writing has the potential to lead to a new type of hegemonic discourse, but we'd like to suggest that often resistance writing fails to account for the realities of limited resources and necessary competition; the utopian impulse of resistance writing—that "we all just get along"—is a difficult (if not idealistic) goal to achieve. One of the more important connections that ecocomposition makes is that as writers strive for discursive homeostasis they rewrite what that homeostasis can be, what that environment can be. Writers, when seen ecologically, possess one of the largest ranges of tolerance and the ability to extend that range by continuously acclimating and altering their habitats. Writers who enter Cooper's web are most often subsumed by it, and ultimately, most often acclimate to it in order to survive. Others, however, may also resist the web, shake it, build new threads, start new webs. In other words, like the managerial science of ecology, writing, particularly resistence writing, seeks to manipulate environments for sustaining more organisms who can operate in the exchange of energy with more equity.

Energy exchange is a primary concern for ecologists as all organisms require energy to function. The conventional, simple definition of energy is "the capacity to do work" (Bowers, 13). Work, in turn, can be seen as movement, both on the scale of locomotion and the smaller scale of cellular movement. Work occurs when energy is converted from one form to another. Energy is classified into categories: potential energy, such as the energy stored in muscle tissue, and kinetic energy, the energy which is burned when an object moves—for instance, a pencil which falls from the side of a desk loses kinetic energy through friction with the air or any other molecule with which it comes in contact. Kinetic theories in physics are based on the general understanding that minute particles remain in constant, animated motion. The kinetic theory of gasses contends that the particles which make up gasses move at high speeds in straight lines until they come in contact with (or collide with) other particles; these collisions force sudden changes in velocity and direction and create pressure within the gas. The kinetic theory of heat, likewise, suggests that the temperature of a substance increases with the increase of kinetic energy. The theory of thermodynamics, developed by Alfred J. Lotka, contends that all ecosystems can be "described by a set of equations that represents exchanges of mass and energy among its components" (Ricklefs, 128). In other words, the exchange of energy, how energy passes from organism to organism within a system, can, in fact, map the operation of that system. The frequently used metaphor by which ecologists describe the passing of energy is to question how each organism within a system "earns its living." Again, the economic foundations of ecology are quite clear. Ecologists also speak of consumers and producers, those whose primary role in the system is to provide energy and those who primarily ingest. All organisms pass energy along in one form or another. Basic explanations of food chains depict how various organisms at different trophic levels require certain energy and then pass on certain energy. Aquatic ecologist Raymond Lindeman explained this passing of energy to resemble a "pyramid of energy" in which less energy reaches higher trophic levels. At each level, work performed by organisms requires more than the previous level. A field of prairie grasses for example, in central Florida might stand at the base of a pyramid of energy—we might call this prairie the "primary producer" in an ecosystem. The grasses take in massive amounts of energy from the sun and from nutrients in the water and soil. These grasses reproduce and grow on the cellular level, and they develop into mature, edible plants. Primary consumers, such as white-tailed deer, marsh rabbits, or eastern harvest mice then graze upon the grasses, ingesting the grass' energy and making it their own. Other consumers, then, such as the ticks and parasites that infest many of Florida's deer and rabbits, or coyote (the only large carnivore to persist in the heavily populated areas of northern and central Florida), grey fox, bobcats, or Florida Panther then ingest energy from these herbivoric consumers. In turn, wild boar, racoons, opossum, turkey vultures, and other consumers which eat carrion then ingest energy from those organisms. Others will then gain energy from them. At each passing of energy, some is lost until the consumer at the peak of the pyramid takes in the least amount of energy of organisms at any other trophic level. All organisms are dependent upon the consumption of other organisms at other levels; organisms are interconnected through the economics of energy.

For compositionists, this understanding of the flow of energy through ecosystems is crucial not only because it explains the interconnectedness of systems—indeed the very concept of system implies interconnectedness—but also it explains how knowledge, culture, ideology, and in turn, how discourse moves through systems and how those very systems are reliant upon the passing of discourse in order for the system to survive. That is, like energy, which passes through chemical and biological systems, ideology and knowledge flow through discursive systems. Individuals absorb discursive energy in the form of ideas, knowledge, and ideology through reading and listening, and they in turn transmit this energy to others through their own writing and speaking. This is not to suggest, however, that communication consists of mere transmission and reception of information; quite often, interaction between individuals involves an exchange of energy, communicative interaction, in which the exchange can be seen as symbiotic, or can even result in the generation of new energy, such as new ideas, new insights, and new knowledge. The taste of sugar, for example, is not present in the separate elements of carbon, hydrogen, and oxygen; the essence of sugar emerges only when these separate elements are combined. That is, combined discursive energy often has the potential to give rise to new, "original" knowledge that is more than the sum of its parts. In other words, discourse is both a product and a source of cultural, ideological, and epistemological systems.

Ecology, Nature Writing, and Discourse Studies

We've suggested more than once in this book that ecocomposition is about more than just the study or production of nature writing texts. However, we do recognize that some of the most overt and articulate expressions of the fact that there is a *natural* element to *discourse,* that writing works in ecological ways, come from nature writers. Many nature writers have recognized this relationship between ecology and written discourse. Barry Lopez, for instance, writes in "Searching for Depth in Bonaire," "What I wanted to experience in the water, I realized, was how life on the reef was layered and intertwined. I now had many individual pieces at hand— named images, nouns. How were they related? What were the verbs? Which syntaxes were indigenous to the place?" (24). For Lopez, a writer by trade, his ecological study of Bonaire's reefs and his inquisitiveness about those ecosystems is best expressed in terms familiar to a writer. Just as we have tried to show how ecocomposition can benefit from seeing written discourse through an ecological lense, so, too, does Lopez find benefit in seeing ecology in a writerly light.

Similarly, the first American nature writers were also among the first to recognize the relationships between the production of the written and spoken word and the natural world. Ralph Waldo Emerson, who is arguably the father of American ecological thinking, believed that language functions as a reflection of the natural world, and that the diversity and interconnectedness found in Nature serves as a "vehicle of thought" (Emerson, 816). For the early Transcendentalists like Emerson and Thoreau, there was no separation between words and nature, and their writing suggests that each word and phrase can ultimately be traced to some *original* source on land, sea, or sky. Emerson suggests in "Language," the fourth chapter of his first major work, *Nature,* that "Words are signs of natural facts . . . Every word which is used to express a moral or intellectual fact, if traced to its root, is found to be borrowed from some material appearance" (816). For Emerson, and for the multitude of nature writers following him, the production of discourse and the diversity of the natural world are inseparable. When Emerson wrote "How much finer things are in composition than alone" he was certainly not speaking of the discipline, but his words echo an attention to the acts of assembly and production, urging us to recognize that the production of language or art, like nature, works best with an acute awareness of that very act and the ways in which that act is connected with other acts of composing. More than one hundred years later, Gary Snyder uses ecological ideas to reflect on the composing process.

> Words are used as signs, arbitrary and temporary, even as language reflects (and informs) the shifting values of the peoples whose minds it inhabits and glides through. We have faith in "meaning" the way we might believe in

wolverines—putting trust in the occasional reports of others or on the authority of once seeing a pelt. But it is sometimes worth tracking those tricksters back. ("The Etiquette of Freedom," 1990)

Although Snyder only hints at ecology—wolverines inherently represent predation and natural selection to many of us—and makes only a metaphorical comparison between words and the diversity of life, their connection is at the same time real, tangible, and *natural*.

Bridging Ecology and Composition

In this chapter, we've tried to show that there are palpable connections between the study of natural living systems—ecology—and our own discipline's study of language and discourse—composition. We feel that it is important, perhaps crucial, for ecocompositionists to know something of ecology if we are to successfully teach and work within the borders of a subdiscipline that draws a part of its name from that very discipline. More important, there is much that compositionists can learn about discourse and writing by turning to theories of ecology. After all, both disciplines, composition and ecology, share a common interest in communities and the ways in which their members work in accord with and in opposition to each other. Both look to the environments in which these interactions take place, whether they are biospheres, neighborhoods, classrooms, or texts. Both composition and ecology are concerned with the ways in which various groups and species compete for territory and resources within particular ecosystems. While compositionists speak of subaltern, marginalized groups and their struggles for the right to speak and act autonomously, ecologists speak of the struggles of particular plants and animals and the ways that they adapt and often overcome environmental competition. Both disciplines recognize that environment is of great importance in the construction of identity; plants and animals react and adapt to their environments, while writers often adopt identities that are often shaped and molded by environmental factors. Even more basically and simply, both disciplines are relatively new arrivals on the larger intellectual horizon, but both deal with subjects that are as old as history itself: communities and communication.

As we mentioned early in this chapter, we don't profess to be experts on ecology, nor do we suggest that we've made all of the necessary connections between these two disciplines. This fact is precisely *why* we think that ecocomposition is so significant and exciting; there are so many correlations, parallels, and points of departure to be made in the years to come. We hope that this chapter extends ecocomposition's understandings of ecology and paves the way for new, perhaps unimagined investigations of the relationships between discourse studies, ecological studies, and other related disciplines (we'll discuss more of these connections and possibilities in chapter 6). In the next chapters, we turn to what has been the most productive and active topic of discussion within ecocomposition to date: the incorporation of ecological and environmental studies into writing classrooms and courses and the move to see beyond classrooms to larger "public" systems, locations, and environments.

CHAPTER 4

Ecocomposition and Activist Intellectualism

⊱✄⊰

*By confronting "face to face" the separate realm of nature, by becoming aware of
its otherness, the writer implicitly becomes more deeply aware of his or her own
dimensions, limitations of form and understanding, and the processes of
grappling with the unknown.* —Scott Slovic,
"Nature Writing and Environmental Psychology:
The Interiority of Outdoor Experience"

*All good writing, everywhere and always, is an act of attention. What most
needs our attention now, I believe, is the great community of land—air and
water and soil and rock, along with all the creatures, human and otherwise, that
share the place. We need to imagine the country anew, no longer as enemy or
property or warehouse or launching pad, no longer as homeland to be recalled
from a distance, but as our present and future home, a dwelling place to be cared
for on behalf of all beings for all time.* —Scott Russell Sanders,
Writing from the Center

Public discourse is a necessary element in the democratic process.
—James A. Berlin,
Writing Instruction in Nineteenth-Century American Colleges

*Writing teachers need to relocate the where of composition instruction outside
the academic classroom because the classroom does not and cannot offer students
real rhetorical situations in which to understand writing as social action.*
—Paul Heilker, "Rhetoric Made Real:
Civic Discourse and Writing Beyond the Curriculum"

Coinciding with the move to incorporate ecological and environmental
studies in American universities, a move to make intellectual conversations

more accessible to public audiences has become prolific. Scholars and teachers in nearly every discipline have become interested in moving beyond the confines and imperatives of their particular disciplines, and many educators hope to extend their influence beyond the boundaries of the academy itself. The discussions surrounding this topic pursue new ways to transcend narrow disciplinary focuses, and consequently, they suggest new public roles for intellectuals. This new endeavor—which some scholars have labeled "public intellectualism"—addresses a collective desire for purposeful and constructive discourse about the political and social matters that affect our lives and the world(s) we all share. Within composition studies, these discussions have examined the role of composition theorists, scholars, and teachers in society and the degree to which they are capable of bringing about progressive social change through their own actions. In other words, some compositionists have begun to ask if any or all of us might become public intellectuals; they have questioned whether or not there is anything particular about our discipline that might make us more viable candidates for this role than intellectuals working in other disciplines. While there is little agreement on whether or not and to what degree compositionists might become public intellectuals per se, we argue that despite disagreements over what public intellectualism might be (by definition), some form of activism is (and must be) implicit in ecocomposition. In other words, we contend that ecocomposition's move to include larger public issues, discourses, and audiences within and outside of academic conversations is more than admirable, it is crucial. Moreover, as we shall explain, political and social activism is inherent in ecocomposition; one can no more separate activism from ecocomposition than one can separate hydrogen and oxygen atoms from a molecule of water and still claim that one maintains hold of water. To do so changes its very essence. In other words, one of ecocomposition's very reasons for being is to inquire into ways to bring about political, social, and/or environmental change—both practical, theoretical, and epistemological.

In a 1997 interview with Sid, Michael Eric Dyson—one of many who have been given the title of "public intellectual"—argued that the move from the academic to the public must be a self-conscious decision. He says that the move to public spheres beyond the academy (a sphere he insists is part of the public) is a difficult transition. For Dyson, the role of the public intellectual is that of the "paid pest," one who is paid to carry the work of the academic intellectual to public debates in order to disrupt and make visible certain aspects of those debates. This definition of public intellectual as an academic who moves into other public arenas seems to be a standard understanding of what public intellectuals do. Others, like bell hooks, have identified that the role of the public intellectual is to bring the learning of the academy to mass publics in the languages that reach those audiences. What is coupled with this definition of the public intellectual is a prevailing sense of fame. That is, in order for public intellectuals to function as paid pests, they must be visible speakers or writers who reach large audiences. The unheard, unseen (by a large audience) public intellectual is only an intellectual. What we hope to do here is show that the role of the public intellectual, often reserved for a few highly visible scholars, creates a definition of the public intellectual which places public intellectuals in the cult of celebrity. While such a definition is limiting and reserves the role of public intellectual for only a few, we do not see a need for redefinition of the positioning or action of public intellectuals. Instead, we wish to argue for activist intellectuals—intellectuals who take their work to the streets, as it were, in smaller, more localized public venues. That is, we wish to show how scholars can be a different type of public intellectual and how ecocompositionists must be activist intellectuals.

Public Writing

While the move toward some forms of public intellectualism has pervaded the academy in the latter half of the 1990s and beyond, an equally compelling and related line of thought has emerged within composition studies in particular: the investigation of *public writing* as a significant and important goal of university writing courses.[1] Like the term *public intellectualism,* many of the perspectives regarding this topic have been quite diverse, and they reflect a number of goals, perspectives, and epistemologies. Briefly stated, public writing can be defined as any written discourse that attempts to address an issue of importance to any local, regional, or national group or groups in order to bring about progressive societal change. While public writing does not necessarily have to result in immediate or widespread reform, it always aims at least to sway or influence public opinion in order to stimulate further discussion. Public writing is democratic, as it intends to use the written and spoken word to improve the lives of marginalized or subjugated groups or individuals. For ecocompositionists, this goal has extended beyond a strictly anthropocentric perspective, and much of the public writing generated or facilitated by ecocompositionists has attempted to improve and/or protect the condition of any number of living beings, ecosystems, or natural environments. Public writing in ecocomposition develops partly from the ecofeminist perspective that all oppressions grow from similar ideological places and in order to change one, all oppressions must be addressed.

Not surprisingly, most discussions of public writing in composition have focused on students, particularly those that compositionists encounter in writing courses. While composition theory has developed in sophisticated and meaningful directions over the past forty years, it has, unlike theory in some other disciplines, nearly always evolved in an effort to make the work done in writing courses more meaningful and productive.[2] Composition is unique in that it revolves not around a particular body of knowledge or an agreed upon disciplinary boundary, but around the common goals of first helping students improve their writing abilities, and second, helping students use writing to improve their lives and the lives of others—including the nonhuman others that occupy much of the attention of ecocompositionists. The project of knowledge-making in composition is deeply implicated in how teaching practices are formed and argued for. As Joseph Harris explains in *A Teaching Subject,* "composition is the only part of English studies which is commonly defined not in relation to a subject *outside* of the academy (to literature, for example, or to culture or language) but by its position *within* the curriculum—by its close involvement with the gatekeeping first-year course in writing" (xi). The desire for effective public writing, particularly as it is invested in students, attempts to make our teaching practices more important and worthwhile. For many compositionists, the classroom—or more specifically, the writing course—has emerged as a microcosm of the public sphere, as our point of contact with the "real" world out there somewhere.

In effect, the move toward public writing, like many other important investigations in composition, is a move toward ecocomposition, in that both envision the sites in which discourse emanates from and enters into as crucial. Public writing foregrounds the locations in which discourse occurs, paying particular attention to the constraints and generic qualities of discourse in those locations. In other words, both public writing and ecocomposition share a common interest in *places* and the relationship of discourse in and upon those places. Unlike earlier theories that locate knowledge and discourse within the individual or group, theories of both public writing and ecocomposition count the locations of discourse as vitally important. For instance, Julie Drew argues that compositionists do not have to redirect their attention from their everyday writing and teaching environments in order to "think about the ways in which *place* plays a role in producing texts, and how such relationships affect the discursive work that writers attempt from within the university. In fact, the very idea of nature, or natural environment, in the composition classroom might arguably be subsumed within a larger notion of place that certainly includes," what Drew refers to as *"the place of writing instruction itself"* (58). For Drew, the place where writing instruction occurs is "almost exclusively understood by compositionists as the writing classroom" and "has been and continues to be examined by scholars in the field as a politicized space, and in that sense a politics of place is by no means a new idea." Drew contends that "the politics of the classroom as embodied location is informed in multiple and conflicting ways by other places in which discursive pedagogics also abound" (60).

While the sites of public writing for many compositionists are primarily human-centered—as the notion of the public sphere for many brings to mind the town square, city hall, or the local newspaper, if not the very classroom in which composition is taught—public writing can both emerge from an array of sites and locations and effect those very sites and all that live in and depend upon those sites. Moreover, even when the arenas of public discourse exist in human-constructed environments, the issues that public discourse address often account for and recognize nonhuman others. Recently, environmental concerns have become a predominant part of many public discourses. We offer but a few examples: the recent growth of environmental publications such as local event newsletters, the increase in outdoor and environmentalist magazines and the number of environmentalist/conservationist issues addressed in outdoor magazines such as *Environment Hawaii, Florida Sportsman, Making Waves* (a publication of the Surfrider Foundation), and even in magazines with international readerships, such as *National Geographic, Sports Afield,* and *Outside Magazine;* the increasing number of debates between political candidates who proclaim their recognition of the importance of new environmentally sound legislation such as George Bush who claimed to be the "environmentalist president" or more publically visible secretaries of the interior James Watt, Bruce Babbitt, or Gale Norton; public congregations like the 1992 Earth Summit in Rio de Janeiro, or 1989 Basle Convention during which an international treaty was ratified to limit the dumping of waste in Third World countries; the pervasive use of "environmentally friendly" advertising to sell products and the often misleading information disseminated in the name of consumerism, such as the rash of advertisements depicting sport utility vehicles as the means to "get back to nature"; listservs and other on-line discussions about environmental and ecological subjects both inside and outside of academia; the increase in ecothriller writers since Edward Abbey's *The Monkey Wrench Gang* and John D. MacDonald's Travis McGee series; and the increase of selections from environmental "debates" included in first-year writing texts, both rhetorics and readers (obviously, we could go on and on here). If we believe that power is entrenched in discourse and that language is an instrumental tool in shaping knowledge and reality, we could, by extension, assume that public writing can have tangible, immediate implications in the world and these examples seem to suggest that there is an ecology of publically writing and talking about environment which, in turn, effects a variety of publics and environments.

Rhetoricians and compositionists have turned toward public writing for a number of reasons. First and foremost, such an approach gives student writing real significance; public writing often allows students to produce meaningful discourse that has the potential to change their lives and the lives of others. In this respect, students see public writing as more "real" than, for example, an essay about what they did last summer or an analysis of a particular piece of literature—writing assignments that students recognize as having a singular audience of the instructor. Public writing can help students to see the value of adopting a particular rhetorical stance, since public writing is often directed toward a particular audience that might be influenced by the student's writing. Students often come away from a course or assignment that focuses on public writing with a better understanding of the importance of shaping the style, form, and tone of their written work in ways that might be most persuasive and compelling. In addition, public writing more easily allows students to see that language is a powerful tool for affecting change in the world. When a student's writing generates further public discussion or leads to some societal change, he or she comes to see how discourse is deeply implicated in the structures of power in a society. Obviously, public writing of this sort is something we'd all like to realize in the writing courses we teach. Unfortunately, few compositionists know where "the public" is located, and even fewer have thought in depth about what public writing might entail beyond letters to the editor of a newspaper or to their local congressman. Quite frequently, students are asked to write letters to the editor about some issue of local importance or of interest to the student so as to test the waters of public writing. Such an exercise is thought to offer students the opportunity to face "real" rhetorical situations and to be heard publically. However, such limited approaches to public writing are extremely problematic. First, because of the brevity of most letters to the editor, students gain little experience writing, making rhetorical choices, inventing, revising, composing. Second, only a small percentage of the letters will actually be read by an audience larger than the newspaper's editorial editor; this simply extends the often singular audience of the teacher to an audience of two. In addition, even if such letters are published and reach a larger audience, they are most likely to read by the audience with an err to the absurd. Let's face it, when we read letters to the editor, unless the author's name is recognizable or noted with a qualifying affiliation, we are likely to dismiss the claims of the author. Credibility is difficult to establish in a letter to the editor; the forum is often regarded as a sounding board for local loud mouths with opinions about just about anything. More important, however, is the artificial construction of "public writing" that the letter to the editor assignment creates. In the same way that spending a weekend in an air-conditioned hotel room in a rainforest, watching a video presentation about the what one might see there, and taking a quick guided tour along a boardwalk path doesn't provide an accurate experience of a rainforest; it provides a simulation of the experience (granted, simulation is a *kind* of experience). So too does the letter to the editor assignment provide only a quick, shallow tour of public writing.[3] In such an experience, the tourist has been guided to a safe, predictable version of a wild place. Such writing assignments are simply safe tours of the public, a park or zoo glimpse of a sphere that is much more dynamic than a letter to the editor. Such assignments often protect students from finding out that there are other organisms out there that might bite, that where one steps, what one does draws recognition, and sometimes opposition, from the environment into which one enters discursively. Cooper's web identifies that part of entering into public discourse is finding out how the web reacts to us as well as how we may react to it. The letter to the editor stabilizes the web, and is in effect like walking along a short paved path through a small portion of a rainforest and claiming that one has "experienced" a rainforest.

Even among the scholars and theorists in composition who have written about public writing there is little consensus on how we might envision or further theorize public writing. Their conceptions of public writing are as diverse as the public itself. To some composition theorists like S. Michael Halloran and Lester Faigley, public writing necessitates a reorientation to public discourse in the ancient rhetorical tradition and an orientation to a "rhetoric of citizenship" within the writing classroom. Halloran's 1982 "Rhetoric in the American College Curriculum: The Decline of Public Discourse" was among the first to rightly assert that "we might well turn some of our attention to the discourse of public life" and in the process enable students to see themselves as "members of a body politic in which they have responsibility to form judgements and influence the judgements of others on public issues" (263–64). Similarly, Faigley's *Fragments of Rationality* suggests that the newer political awareness in composition studies has the potential to "recover a lost tradition of rhetoric in public life" (71). Notice the concepts of public life and members of larger systems. What Halloran and Faigley are, in fact, suggesting is that we resituate discourse within ecological frameworks, that we look at larger rhetorical systems. For other compositionists, like Joseph Harris, the term "public" is a useful metaphor for how we might imagine the writing classroom in the 1990s and beyond. Harris argues for a "view of the classroom as a public space rather than as a kind of entry point into some imagined community of academic discourse" (109). That is, Harris, like Drew, envisions the classroom not as separate from the public, but an ecosphere of its own which reacts with and is part of other publics. Other theorists, including Susan Wells and Irene Ward, have examined composition's desire for efficacious public writing—especially as it is invested in students—and have theorized ways to construct public spheres that students might enter through discourse. Wells' "What Do We Want from Public Writing" is perhaps the most thorough investigation of public writing; in it, she posits: "If we want more for our students than the ability to defend themselves in bureaucratic settings, we are imagining them in a public role, imagining a public space they could enter" (326). Ward's "How Democratic Can We Get?: The Internet, the Public Sphere, and Public Discourse" discusses the potential for "the Internet to become a public sphere, and, hence, a forum that private individuals could use to democratically influence the state" (365). It is interesting and useful for our conversation here to note that Wells and Ward theorize the public sphere as a complex, multilayered discursive space: an intricate discursive ecosystem comprised of many smaller, more specialized discursive ecosystems (though they never use the word "ecosystem" specifically). Still others have attempted to introduce new forms of public writing into their courses through "service learning." Composition-ists like Bruce Herzberg suggest that the kind of civic participation that service learning courses in composition offer can allow students to discover "real applications [for] their knowledge in the organizations they serve" and also learn that they "can use their knowledge not only to get jobs for themselves but also to help others" (308). Similarly, Ellen Cushman has argued that composition courses can increase participation in public dis-course by "bridging the university and community through activism" (7).

The service-learning movement has itself become an important conver-sation in composition studies, so much so that in 1999 CCCC chair Victor Villanueva appointed a Service-Learning Committee to encourage "teach-ing and research initiatives involving service learning and will oversee a small grant that NCTE received" (*Committee Cronicle* 9:6). This committee is developing a website for resources on service learning.[4] Likewise, collec-tions like Linda Adler-Kassner, Robert Crooks, and Ann Watters' *Writing the Community: Concepts and Models for Service-Learning in Composition*, have begun to explore the role of service-learning in the composition class-room. For Linda Adler-Kassner, Robert Crooks, and Ann Watters, the service-learning "revolution" has become crucial specifically because "in the growing number of schools where service learning has been implemented, either on a course-by-course or programatic basis, both faculty and student participants report radical transformations of their experiences and under-standing of education and its relation to communities outside the campus" (1). The authors continue to note that few compositionists, however, know much about service learning and that conferences such as CCCC have been slow to accept panels on the subject and that MLA has accepted none. For ecocompositionists, like Annie Merril Ingram, service learning must be a central part of ecocomposition. In her essay, "Service Learning and Eco-composition: Developing Sustainable Practices through Inter- and Extra-disciplinarity," Ingram contends that bringing service-learning components to the ecocomposition classroom "benefits not only students and teacher—in terms of greater motivation, productivity, and investment in the course—but also the wider community, while broadening the scope of contribution of the class in real tangible ways" (209). We will return to notions of service learning in the ecocomposition classroom in our discussion of pedagogy in the next chapter, but for now what we want to acknowledge is the growing concern for both public writing and the manifestation of that interest in the service-learning movement.

These and many other compositionists see public writing as among the most significant topics in composition studies today. Although there is little agreement as to how a composition course, focusing on public writing might be enacted, we think it is safe to say that all of the compositionists just mentioned are working toward the same goal—despite varying degrees of epistemological and methodological differences. While most of the discussions concerning "publicness" in composition studies have focused on the discourse that students produce, some of these discussions have examined the broader issue of the role of composition theorists, scholars, and teachers in society and the degree to which they are capable of bringing about progressive social change through their own actions. In other words, some compositionists have begun to ask if we might become *public intellectuals,* and they have questioned whether or not there is anything particular about our discipline that might make us more viable candidates for this role than intellectuals working in other disciplines. Stanley Fish, the noted literary critic and theorist, suggests that public intellectualism is not possible in English studies (or any other discipline, for that matter). In our attempts to more fully theorize public intellectualism and activism, we've found it useful to draw upon the work of noted social and cultural theorists including Jürgen Habermas, Oskar Negt, Alexander Kluge, and (particularly) Nancy Fraser, who have provided various careful studies of the concept of "the public." By doing so, we develop a more holistic, sophisticated analysis of the subject of public intellectualism in an effort to explain exactly how and to what degree ecocompositionists might become more successful in their attempts to work not only toward a more democratic society but also a more ecologically sustainable society. We explore the larger implications of incorporating ecological concerns—and by this we mean all of the useful meanings that we've already discussed, including ecology as a theoretical framework, ecology as a metaphor for how discourse is produced and consumed, and ecology as a holistic conceptualization of the Earth and its inhabitants—into various public and academic conversations. In the process, we describe a new approach to thinking about the roles that ecocompositionists might assume in their attempts to enact progressive political and social change.

Stanley Fish, Jürgen Habermas, and the Question of the Public Intellectual

Certainly, the urge to make what we do in English studies more culturally relevant is of concern to many members of the discipline. In fact, the desire to influence public opinion and bring about progressive societal change is of interest to scholars and intellectuals in nearly every discipline. Many of us have begun to ask what we can do to help bring about changes in society through public forums. Scholarly journals and edited collections in English studies are filled with essays by authors who question whether or not intellectuals are capable of moving beyond disciplinary and academic borders to engage in deliberation and debate with larger public audiences. In English studies, scholars such as Michael Bèrubè, Cary Nelson, Bruce Robbins, and others have enabled us to reconsider our roles within English departments and to recognize the political functions of academic work and the ways that universities can serve as vehicles for activism. In composition, scholars including Ellen Cushman and Patricia Bizzell have put forth interesting and important investigations of the particular role of compositionists in bringing about change in both the academy and in society as a whole. Those interested in developing ecologically sound curriculum such as C. A. Bowers and Frederick O. Waage and ecocompositionists in particular such as Randall Roorda and Derek Owens have questioned whether or not we might promote environmentalist perspectives through and alongside our positions in the academy. In short, the desire to become a public intellectual, one who speaks to diverse audiences on issues that affect segments of society outside of academia, has been discussed from a variety of perspectives, and some scholars have put forth some careful and constructive strategies for moving toward public intellectualism.

However, despite our inclinations, bringing about progressive societal change is not easy for intellectuals, or for anyone else for that matter. It is quite difficult to enter into public debate in nearly any form, and ecocompositionists are no better equipped to deal with the complex array of forces in the public sphere than are other members of society. As some scholars have noted, our jobs in the academy do not necessarily contain a democratic element, and to do those jobs effectively often leaves us with little time for activism outside of the academy. Stanley Fish, for example, suggests that academic work has little effect on public issues and opinions and that only those who have the frequent attention of vast segments of a society can be considered effective public intellectuals. In order to fully understand Fish's perspective on the question of the public intellectual, it is necessary first to understand how he defines the subject. In *Professional Correctness: Literary Studies and Political Change,* Fish offers the following definition:

> A public intellectual is someone who takes as his or her subject matters of public concern, and *has the public's attention.* Since one cannot gain that attention from the stage of the academy (except by some happy contingency), academics, by definition, are not candidates for the role of the public intellectual. Whatever the answer to the question "How does one get to be a public intellectual?" we know it won't be "by joining the academy." (1995, 118)

Based upon this definition, it is easy to see why he believes that public intellectualism is not available to most members of the academy. Certainly, we must agree with him that we cannot gain the attention of "the public" from the stage of the academy, and, if we accept this definition of public intellectualism, we must agree with Fish that it is not possible to assume this role while maintaining our positions in the academy. To be fair, Fish works from a long tradition that defines the public intellectual as someone who speaks to the masses on matters of general interest. In this respect, he is correct in suggesting that this sort of public intellectual does not exist in our society at present. Thinkers like Emerson or Einstein, who were seen as sages on any number of intellectual issues, are few and far between today—particularly in the United States. In the postmodern world, society no longer turns to intellectuals for guidance on broad social and political issues. We see this reflected in popular films and media in which the solution to almost any problem used to be to "call the professor."[5] Contemporary films rarely turn to the intellectual for solutions or input (usually they call in Schwarzenegger or Van Damme, but that's another story). However, what we would like to suggest is that even though public intellectuals in the traditional sense— or film portrayal of that tradition—no longer exist (and perhaps rightly so), ecocompositionists who are interested in participating in social and political debates in society still have a number of options available to them. Effectively exercising these options requires that we go beyond the traditional sense of public intellectual work. There are, of course, many problems with Fish's definitions and assumptions which allow him to see "the university" as separate from "the public." However, ecocomposition sees the university as the public, all part of the same system, all the same place.

Interestingly, Fish's definition of a public intellectual seems to contain many of the same conceptions as those forwarded by Jürgen Habermas. In *The Structural Transformation of the Public Sphere,* Habermas offers an analysis of the bourgeois public sphere as a model for how forms of the public sphere might be formed in the present and future. Where Habermas examines the location of public discourse—labeled "the public sphere" by many theorists—Fish examines the degree to which modern-day intellectuals might enter and influence the discourse produced in this sphere. Both Habermas and Fish suggest that today it is very difficult to elicit democratic political and social change through public discourse. Habermas suggests that the public sphere has eroded, and its erosion can be "clearly documented with regard to the transformation of the public sphere's preeminent institution, the press" (181). He concludes that the bourgeois public sphere disappeared with the emergence of "welfare-state mass democracy" (184). In other words, Habermas suggests that for citizens today, the public sphere no longer exists, and, consequently, public discourse does not exist in a form that might enable citizens to bring about progressive social change. While Fish does not examine the history of the bourgeois public sphere, he does suggest that there was a time when intellectuals had the opportunity to engage in public discourse for the purpose of bringing about changes in society or at least swaying public opinion.

> As things stand now, the public does not look to academics for *general* wisdom, in part because (as is often complained) academics are not trained to speak on everything, only on particular things, but more importantly because academics do not have a stage or pulpit from which their pronouncements, should they be so inclined to make them, could be broadcast . . . I say "as things stand now" because academics, or at least a visible number of them, once did have such a pulpit, the college president or major deanship, offices that for a long time carried with them not only the possibility but the obligation of addressing issues of public concern. (119–20)

Clearly, both Habermas and Fish look to the past, seeing a time when an arena for public discourse (be it a sphere or a pulpit) did indeed exist. Despite the fact that they differ on their conceptions of how these arenas disappeared, both suggest that the true public sphere in the sense that they write of does not exist in society today.

In addition, both Habermas and Fish share a number of other conceptions regarding the location and activity of public discourse. Both, for example, suggest that public discourse is only worthwhile if it reaches a large segment of the population who are able to act upon it in some way.

Furthermore, they both seem to suggest that public discourse must address an eclectic audience, since speaking to an assorted constituency is the only way to bring about widespread changes in thinking and practice. In other words, they assume that public discourse must address the "general public," and the term public is often taken to encompass all members of a society, or at least a representative microcosm of them. Of course, such a limiting view of "the public" denies opportunities for public discourse to reach and effect smaller, local audiences. For instance, speaking or writing about local referendums on water quality would not be affective under this definition of public, or if some affect is managed, then it has been through something other than public intellectualism. However, for ecocomposition, such local public participation is crucial and necessary in a variety of public arenas.

Habermas' account of the public sphere stresses the singularity of the bourgeois conception of the public sphere, its claim to be *the* public arena, in the singular. Habermas asserts that the bourgeoisie in seventeenth- and eighteenth-century Europe conceived of "the public sphere as something properly theirs" (24). His narrative seems to agree with this conception, since it "casts the emergence of additional publics as a late development signaling fragmentation and decline" (Fraser 122). That is, Habermas seems to suggest that any departure from this conception of a singular public sphere is a departure from the ideal. Habermas' narrative is based upon the underlying assumption that confining public discourse to a single, overarching public sphere is a desirable and positive move, whereas the proliferation of discourse in a multiplicity of public spaces represents a departure from, rather than an advance toward, democracy. Ecocomposition, as we have suggested, must see many public spheres/ecosystems and must participate in them. That is, just as ecologists understand that action within smaller ecosystems will ultimately effect larger ecospheres and biospheres and that the actions of organisms within their range of tolerance will have effect on other organisms that may carry these affects to other ecosystems and ultimately may spread them throughout the biosphere. Similarly, ecocompositionists begin to see that local activity directed at more confined publics may begin to effect larger publics as the affect of local activism disperses throughout a range of discursive environments.

Fish seems to argue that public discourse must always address a single public consisting of all or many members in a society. He asks the question, "What do we say to the *public,* that generalized body that wants, not unreasonably, to believe that the cultural activities it sustains have a benign relationship to its concerns and values" (115–16). There are several aspects of this question that parallel Habermas' conception of the public sphere.

First, Fish asks "What do we say to 'the *public,*'" a question that seems to suggest that there is just one singular group of individuals to whom intellectuals (or ecocompositionists) might speak. Moreover, he emphasizes this point by referring to the public as a "generalized body"—a definition that seems to exclude the sort of subaltern or counterpublics that Negt, Kluge, and Fraser discuss (more on this later). Finally, Fish suggests that the topics of public discourse must be confined to matters of common concern, since he appears to be most mindful of the public's "concerns and values." Fish reinforces this belief that public discourse must be of common concern when he argues that the public intellectual is "someone to whom the public regularly looks for illumination on any number of (indeed all) issues" and makes his point even clearer when he suggests that the public "does not look to academics for this *general* wisdom" (119).

Quite simply, Habermas and Fish suggest that the only discourse truly worthy of being called *public* discourse (and similarly, the only intellectual worthy of being called a public intellectual) is that which addresses a large segment of society comprised of a diverse cross section of the general population and speaks about matters that are of concern to all or most of these assorted individuals. While all of these conceptions hold true in the traditional sense, we'd like to suggest some ways that ecocompositionists can move beyond the traditional sense of what it means to be a public intellectual. In fact, we would argue that it is necessary for us to rethink what it means to be an intellectual working in the public sphere today, and this rethinking requires us to take into account the particularities of the postmodern world in which we live and work. We can, and should, move beyond the longstanding conventional definition of public intellectualism if we are to influence public decision-making and action. For this reason, we adopt the term "activist intellectual" to distinguish our version of public intellectualism from the traditional definitions.

Rethinking Public Activism

At this point, we'd like to turn to several of the most significant points that Nancy Fraser makes concerning the public sphere. Fraser questions many of the assumptions that Habermas makes in *The Structural Transformation,* and, to our thinking, offers a more critical and useful stance on the role citizens might play in public spheres. In "Rethinking the Public Sphere," she makes several suggestions toward a new conception of discourse in public spheres. Interestingly, many of these same suggestions apply to the role of intellectuals in general, and ecocompositionists in particular, in public spheres. While we are not proposing that all of Fraser's suggestions refute what Fish has to say, we are proposing that much of Fraser's work is pertinent and applicable to the question of the public intellectual.

There are a number of suggestions that Fraser makes regarding public spheres that can be useful in rethinking how intellectuals might have greater political, social, and ecological import. For example, her analysis of the sites in which public discourse (and by extension, public intellectualism) can occur is quite compelling. Counter to both Habermas' and Fish's confidence in an all-encompassing site for public discourse that is comprised of a cross section of society, Fraser contends that in stratified societies, "arrangements that accommodate contestation among a plurality of competing publics better promote the ideal of participatory parity than does a single, comprehensive, overarching public" (122). She suggests that in societies whose basic structure generates unequal social groups in relations of dominance and subordination (as is the case in the United States), full parity of participation in public discourse is not feasible. Despite the fact that all members of a society may be *allowed* to participate in public discourse, it is impossible to insulate special discursive arenas from the effects of societal inequality. This being the case, she goes on to assert that the disadvantages that certain groups face are only exacerbated where there is only one single arena for public discourse. If there were only one site for public discourse, members of subordinate groups would have no arenas for deliberation among themselves about their needs, objectives, and strategies. They would have no "venues in which to undertake communicative processes that were not, as it were, under the supervision of dominant groups" (123). In other words, if there were only a single public sphere, subaltern groups would have no discursive spaces in which to deliberate free of oppression. While we're not suggesting that intellectuals are marginalized per se, in many circumstances, they speak from and for marginalized viewpoints. In composition's investigations of the critical categories of race, class, and gender, much attention has been paid to the examination of the ways that certain groups have been excluded from various features of dominant society, and much of our public discourse would and does address these inequalities. Ecocompositionists are among those who are most interested in moving beyond academic discourse and participating more fully in social and political debates as public intellectuals. We think it's fair to say that the perspectives of these scholars and teachers are often excluded from the larger public discussions in, for example, the media. Speaking for and about environments—a central feature of ecocomposition—is often seen as a radical move, and such endeavors into public discourse are often dismissed as leftist, tree-hugging maneuvers that are hardly worth the attention of society as a whole. The portrayal of "green" academics in the media is hardly flattering.

Fraser suggests that it is advantageous for subordinated groups to constitute alternative sites of public discourse—what she calls "subaltern counterpublics." She envisions these sites as "parallel discursive arenas where members of subordinated social groups invent and circulate counterdiscourses to formulate oppositional interpretations of their identities, interests, and needs" (123). Fraser sites for example the late twentieth-century U.S. feminist subaltern counterpublic, with its diverse array of journals, bookstores, publishing companies, film and video distribution networks, lecture series, research centers, academic programs, conferences, conventions, festivals, and local meeting places (123). While subaltern counterpublics are not always inherently democratic or progressive, they do emerge in response to exclusions and omissions within dominant publics, and as such they help expand discursive space. In general, the proliferation of subaltern counterpublics means a widening of discursive contestation, and "that is a good thing in stratified societies" (124). Subaltern counterpublics can serve at least two functions. On the one hand, they function as spaces where oppressed or marginalized others can withdraw, regroup, and heal; and on the other hand, they function as "training grounds" and bases for the development of discourse or action which might agitate or disrupt wider publics.

Ecocompositionists can take part in creating such counterpublics, and we must also look for alternative sites in which to voice our opinions on social and political issues. Outlets such as the ASLE listserv, the variety of presses now publishing scholarship about environmental and ecological issues,[6] columns, and articles in more popular publications such as magazines and newsletters, and talks and papers given at local activist meetings all provide outlets for ecocompositionists to use their training as writers and rhetors to be heard publically. Obviously, public discourse need not be limited to a single discursive arena that reaches a huge cross section of society. Ecocompositionists must look to the variety of spaces already available to us for work as activist intellectuals. For example, we can promote change in our communities and public spheres through three general and interconnected means: through the classroom, through scholarship, and through our own public actions. These means must reinforce each other if our activism is to have any effect at all. That is, text must function ecologically, in relation to, as part of a larger system. Additionally, we must not make the mistake of assuming that our individual work will bring about sweeping changes in the thoughts or actions of society or the public as a whole. Even our use of the term "public" is problematic, as the public is a space filled with contradictory voices, varying opinions, and separate discourses in an endless number of combinations. Ecocomposition itself is a contested space, drawing a variety of definitions, with different participants setting different agendas, just as are composition studies and English studies. But what is important to note is that our work can dramatically affect the individual lives of those with whom we come into contact (human and nonhuman) and can have valuable and significant effects in our communities in general—as long as we do not assume that these changes will occur overnight or that they will necessarily affect society as a whole. This is one reason, as we explain in the next chapter, that ecocomposition pedagogies must not only be active, but they must be active locally.

We might resist looking at social change as exisiting only on a wider spectrum, for in doing this we may come to feel that social and political forces are unalterable in the present, and we might be forced to agree with Fish in saying that a time when academics are capable of making social change "is not coming soon, and I do not feel that anything you or I could do will bring it closer" (2). Instead, we might begin to view social/ecological change on a more immediate and intimate level of interaction and allow it to fit into our particular situations and communities. We might begin to conceive of ourselves as activist intellectuals, individuals who work through, around, and beside our academic occupations to bring about social reform on local levels. The situation for the ecocompositionist who wishes to have some social agency is not black and white; we can choose to work with the smaller communities and individuals we come into contact with in our everyday lives. Rather than looking for immediate, widespread results that we ourselves have brought about individually, we might strive for cumulative change brought about by many individuals over extended periods of time. This particular aspect might seem glaringly obvious to many ecocompositionists. For many, their work in the public sphere already exists through smaller counterpublics. That is, to borrow from the cliched bumper sticker, we must think of global publics but act within local environments.

Ecocompositionists often spend much of their time outside of the classroom working for environmental or other activist organizations (those who don't, should), and nearly all of these began as small, grassroots organizations striving to bring together under and unrepresented voices and perspective with the goal of preserving or protecting particular habitats or species. Many also write about or speak about issues publically, but most important is that ecocompositionists get their hands dirty in the local mud. Of the thousands of environmental organizations today, most work from and for specific local or regional locations. Even such giants as Greenpeace and Sierra Club pale in comparison to huge purveyors of public opinion such as CNN and other media conglomerates. And although smaller organizations have neither the monetary resources nor the exposure of the mainstream media, they have been responsible for some significant changes in public policy, and, consequently, in environmental protection, preservation, and restoration. The Surfrider Foundation, a nonprofit organization dedicated to the protection of coastal beaches and waters, is an excellent example of an environmental organization that has brought about significant environmental changes through discourse. Founded in 1989 by a small group of Southern California surfers, the foundation has grown to over twenty-five thousand members in 2001, with local chapters throughout the United States, Australia, Brazil, Europe, and Japan. The Surfrider Foundation owes much of its success to their attention to discourse and education. The foundation publishes an award-winning bimonthly magazine, *Making Waves;* designed and distributes *Respect the Beach,* a coastal education packet for K–12 educators; solicits water-testing reports, press releases, and local updates from chapter members, both nationally and internationally; and coauthored the Beaches Environmental Assessment and Coastal Health Act of 2000 (B.E.A.C.H. Bill) which was signed into law by President Bill Clinton. Importantly, the Surfrider Foundation uses the strengths of its members, utilizing a diverse and talented advisory board comprised of biologists, environmental scientists, physicians, professional writers, university educators and administrators, and other experts. The success of grassroots environmental organizations like Surfrider should serve as a reminder that ecocompositionists are capable of various forms of public intellectualism—perhaps best if we can realize this goal *through* our intellectual occupations, not instead of them. It is possible, as Richard Ohmann suggests, to "work where you are" and "to challenge entrenched inequality and the arrogance of power that nearly saturate our main arenas of public discourse and social action" (256). As compositionists trained in rhetoric and writing in various contexts and interested in environmental/ecological issues, we are uniquely positioned to offer our abilities as rhetors and writers to local organizations and to be vocal in local debates.

Fraser also critiques the assumption (which both Habermas and Fish seem to share) that discourse in public spheres should be limited to deliberation about the common good, and that the appearance of private issues and interests is always undesirable. What Habermas and Fish both seem to argue is that public spheres must be sites where private persons deliberate about "public matters." In a sense, they are correct that matters of public debate must affect a number of individuals to be of importance to society as a whole. However, what they fail to recognize is that the term "public" is ambiguous and open to interpretation. Fraser argues that there are several usages of the term "public," and in regard to the sense that the term might mean "of concern to everyone," she suggests that only participants can decide what is of common concern, and there is no guarantee that all of them will agree. Fraser suggests that the term "public" is "ambiguous between what objectively affects or has an impact on everyone as seen from an outsiders perspective, and what is recognized as a matter of common concern by participants" (128–9). Only participants can decide what is of common concern to them. However, there is no guarantee that they will all agree, and what will actually count as a matter of common concern will be decided through discursive contestation. Any consensus that has been reached through such contestation will have been reached "through deliberative processes tainted by the effects of dominance and subordination" (131). In other words, those who are in power get to decide what is a public issue and what is not.

Fraser asserts that the terms "public" and "private" are not simply absolute definitions of the two basic spheres in society; they are "cultural classifications and rhetorical labels" (131). As such, they work ideologically to reinforce the boundaries of public discourse in ways that disadvantage subordinate groups and individuals. For example, the issue of pesticide use on farms was, until the publication of Rachel Carson's *Silent Spring,* considered to be a private matter decided upon by landowners. The dangers of pesticides, if they were seen as dangers at all, were thought to be confined strictly to the land on which pesticides were sprayed. Carson's book raised public awareness of the widespread, long-term effects of pesticide use, and it convinced many readers that pesticides affect the larger ecosystems of which individual farms were just a small part. Carson's book generated worldwide debate on pesticide use and transformed a previously "private" matter into a public one.

While Fish suggests that a public intellectual must "take as his or her subject matters of public concern," we would argue that these concerns need not address matters of interest to all or most of the members in a society. Instead, activist intellectuals might take up matters that address particular groups and their interests. Fish notes a number of intellectuals who take up more specialized causes, but he suggests that they are not able to speak on more than one specific issue and are not, therefore, public intellectuals.

> These [rent for a day intellectuals] and others will only get the call when the particular issue with which they are identified takes center-stage and should that issue lose its sexiness, their media careers will be over. That is why they are "cameo" intellectuals or intellectuals for a day; a public intellectual, on the other hand, is the *public's* intellectual; that is, he or she is someone to

whom the public regularly looks for illumination on any number of (indeed all) issues, and, as things stand now, the public does not look to academics for this *general* wisdom, in part because (as is often complained) academics are not trained to speak on everything, only on particular things. (119)

We would argue that our definition of the activist intellectual need not be confined to those who can speak about all issues to all people, nor to those who have achieved the fame of being able to make cameo appearances in public debates. Instead, we should view activist intellectuals as those who can speak to *any* group outside of the academy on any issues—particular or general. There is nothing wrong with taking up specific issues in public debates, and our professional and personal interests often lead us to these issues. We note, for example, the various environmental concerns addressed in *Green Culture: Environmental Rhetoric in Contemporary America*. Throughout this collection, rhetoricians turn their critical gaze toward environmental issues of local concern as well as more global issues: Steven B. Katz and Carolyn R. Miller on low-level radioactive waste sites in North Carolina; Craig Waddell on the saving the Great Lakes; James G. Gentry on Yellowstone; Zita Ingham on Red Lodge, Montana to name but a few. These rhetorical readings begin a push toward activism, for they stand as examples of how ecocompositionists, rhetoricians might lend their particular expertise to speaking about local issues. Granted, the overriding agenda of *Green Culture* is not only to offer rhetorical analysis of environmental debates, but also to bring together a study of how "the environment" gets written and discussed. One of the key points brought up in *Green Culture* is the idea that when it comes to public issues like environmental debates, it is important to hear from a range of voices on the subject. That is, "the public" involved in such debates must hear from not just a particular kind of intellectual—a scientist, for instance, who brings quantifiable data to a debate—but from many who have expertise in a range of approaches to the debate: locals who are have emotional connection to a place, land rights activists, or rhetoricians who can shed some discursive light on how the argument is being presented. In other words, activism in local environmental debates, must come from more than the cameo intellectual appearance. The presumed omniscient public intellectual should not be the only voice heard; the activist intellectual must also speak out.

As a matter of example of how such activism might occur, we'd like to turn briefly to the 1992 Save Our Sea Life debates in Florida. Briefly put, the S.O.S. debates surrounded a proposed amendment to ban all inshore gillnet fishing in Florida. The debate pitted the commercial fishing industry against the sport fishing industry (billed as the conservationists). This debate—and the subsequent passing of the amendment on November 8,

1994 by an overwhelming 72 percent vote in support—was one of the most highly visible and volatile environmental debates in Florida's history. Calling into question economic concerns, cultural concerns, conservationist concerns, and a host of emotions, this debate was played out for more than two years in the media, at meetings, at boat and angling trade shows, and a host of other discursive sites. The debate was also played out on the bodies of conservationists and commercial fishermen as protests and rallies often turned to brawls over issues such as "don't take away my family's livelihood." In addition, this debate was played out on the bodies of fish and other marine animal populations as frequent wholesale slaughter of fish were conducted by commercial fishermen as protest and warnings to Floridians.[7] Sid, who at the time was writing a short, weekly fishing report for Florida's *The Fisherman* magazine, began speaking and writing about the S.O.S. debates. By no means was his a widely heard voice, but his rhetorical readings of cultural rhetoric, his weekly notes about preserving fish populations, his public talks about the debates, and his use of the debates for rhetorical analysis in his composition classes (both FY and advanced) pushed toward the kind of activist intellectual we envision.

Fraser also questions the assumption that it is possible for interlocutors in any public debate to actually bracket status differentials and to participate in discourse as if all of the members of a public sphere were social equals. Fraser suggests that in any public sphere, it is impossible to effectively bracket social differences among interlocutors.

> But were they [the differences between interlocutors] really effectively bracketed? The revisionist historiography suggests that they were not. Rather, discursive interaction within the bourgeois public sphere was governed by protocols of style and decorum that were themselves correlates and markers of status inequality. These functioned informally to marginalize women and members of the plebeian classes and to prevent them from participating as peers. (119)

In this respect, Fraser is talking about informal impediments to participatory parity that can persist even after everyone is formally and legally licensed to participate. Certainly, there are no legal restrictions on intellectuals participating in public debates in the United States today, regardless of the circumstances. Intellectuals are free to speak on practically any topic in society today, and, as Fish notes, he himself has seen a number of them "peering out from [his] television screen" (118). While there are no legal barriers keeping intellectuals from entering public discourse, there are other more subtle forces that cannot be easily disposed of through legislation. Fraser notes a number of these informal impediments that might keep citizens in general, and intellectuals in particular, from participating in public discourse. She cites, for example, a familiar contemporary example drawn from feminist research. It has been documented that in male and female deliberations, men tend to interrupt women more than women interrupt men; men also tend to speak more than women; and women's interventions are more often ignored or not responded to than men's. Deliberation and the appearance of participatory parity can serve as a mask for domination. Acknowledging the work of theorists like Jane Mansbridge, Fraser argues that various manipulations of discourse are often used to cloak nearly imperceptible modes of control. Fraser argues that

> the transformation of "I" into "we" brought about through political deliberation can easily mask subtle forms of control. Even the language people use as they reason together usually favors one way of seeing things and discourages others. Subordinate groups sometimes cannot find the right voice or words to express their thoughts, and when they do, they discover they are not heard. [They] are silenced, encouraged to keep their wants inchoate, and heard to say "yes" when what they have said is "no." (119)

Similarly, M. Jimmie Killingsworth and Jacqueline S. Palmer contend that

> Following Burke, we can think of this problem in grammatical terms. Any action may be stated in an active voice sentence, the kernel of the group's identifying story: *I (or we) do this.* In the case of intractable problems, the subject position of one group ("we") cannot be filled with members of another group ("you" or "they"). Rhetorical appeals propose enlargements of the *we* category or mergers of two or more categories, with the ultimate goal being the identification of the "global" public with the "local" discourse community. (7–8)

This melding of the "I" voice into the "we" culture can be likened to Cooper's web, in which the writer must move into the web in order to be heard, and frequently is subsumed by the web and the writer's disturbances of the web absorbed and consumed. Many of these insights into ways in which discourse is used to mask domination and imbalances of power can be applied to other kinds of unequal relations, like those based on class or ethnicity. They alert us to the ways in which "social inequalities can infect deliberation, even in the absence of any formal exclusions" (119). In this respect, the bracketing of differences and social inequalities in public discourse cannot actually be enacted, and assuming that it can actually works to the advantage of dominant groups in a public sphere and to the disadvantage of subordinates. In most cases, it would be more appropriate to *unbracket* these inequalities by foregrounding and thematizing them. Doing this would help to eliminate some of the more pernicious uses of discourse in public deliberation. The assumption that public discourse occurs in arenas which can overlook, bracket, or disregard social and cultural differences is counterfactual.

Undoubtedly, these inequalities taint the deliberation of intellectuals who engage in various forms of public discourse, and the effects of these inequalities are certainly obstacles that intellectuals must recognize and negotiate if they are to deliberate successfully in public settings. The ideology of gender, like the ideologies of race and class, constructs and reconstructs all sorts of discursive interaction. To return to Rachel Carson's public work dealing with pesticides, for example, we can see the degree to which her gender influenced the reception of her work. As Al Gore writes in his introduction to *Silent Spring,*

> The attack on Rachel Carson has been compared to the bitter assault on Charles Darwin when he published *The Origin of Species*. Moreover, because Carson was a woman, much of the criticism directed at her played on stereotypes of her sex. Calling her "hysterical" fit the bill exactly. *Time* magazine added the charge that she had used "emotion-fanning words." She was dismissed by others as a "priestess of nature." Her credibility as a scientist was attacked as well: opponents financed the production of propaganda that supposedly refuted her work. It was all part of an intense, well-financed negative campaign, not against a political candidate but against a book and its author. (xvi)

Carson, too, was cognizant of the role gender affected how her work was received. In her 1951 *New York Herald-Tribune* book and author luncheon speech about her landmark book *The Sea Around Us,* Carson acknowledged that "people often seem to be surprised that a woman should have written a book about the sea." "This is especially true," she says "of men."

> Perhaps they have become accustomed to thinking of the more exciting fields of scientific knowledge as exclusively masculine domains. In fact, one of my correspondents not long ago addressed me as "Dear Sir" — explaining that although he knew perfectly well that I was a woman, he simply could not bring himself to acknowledge the fact. (77)

The imbalances of power that exist in nearly every aspect of society are not likely to magically disappear when some or even all of the actors involved are academics. But the awareness of these imbalances is one step toward overcoming them. Compositionists, in particular, seem well-suited to recognize and surmount some of the effects of this inequality. Much of composition scholarship in recent years has examined the role of ideology in shaping conversations, and composition's particular rhetorical awareness might help to foreground situations in which public discourse is being affected by particular social forces. Ecocompositionists should be quick to recognize that social differences shape public discourse in significant ways and the ways that such differences are played out through and on environment. Robert D. Bullard's important book *Dumping in Dixie: Race, Class, and Environmental Quality* demonstrates well how such social differences are linked with environmental concerns. Likewise, ecofeminists have rightly noted the connections between all socially oppressive ideologies and environmental oppression. We also need to continue efforts to establish new theories of public discourse that acknowledge differences in thinking and writing—theories that might make us more astute as activist intellectuals.

Public Teachers

If we accept that activist intellectualism might consist of more than addressing a singular, overarching public sphere, and that our discourse need not be confined to matters of "common concern," we might be able to conceive of a new definition of what it means to be an activist intellectual. Rather than supposing that our activist efforts must occur in just one way, we might begin to see a variety of opportunities for work that influences political and social decision-making and action in society. Our work in the classroom, for example, might be seen as perhaps the most important and effective avenue of political and social change that is available to us. It's unfortunate that some of us involved in the production of knowledge in English studies seem to feel that our work is somehow divorced from the lives, actions, and issues outside of our universities. While recent theories and pedagogies in both literature and composition have powerfully changed how we teach, we have largely "failed to make a convincing case beyond the classroom for the new view[s] of literacy that we profess inside it" (Harris, 324). Our classroom work *must*, then, attempt to help students develop the real skills that they will need to be successful in their lives both inside and outside of the university. By "skills" we mean not only how to write effectively for future classes and careers (although this is certainly important), but also how to make well-informed decisions about the political and social issues that affect them. That is, in line with Mike Rose, we do not wish to reduce writing to a series of measurable skills.

English departments, and compositionists in particular, play a unique and powerful role in the modern university. At nearly all colleges and universities, students are required to complete at least one and usually two semesters of first-year English. What is most significant, as Evan Watkins notes in *Work Time,* is "the appropriateness of the gatekeeper image to the position of English departments, not only in community colleges but throughout the educational system" (5). In reality, nearly everyone who earns an undergraduate degree in the United States has spent at least one year in frequent contact with an English instructor or professor. First-year English is, after all, something close to a rite of passage into the educated middle class. As Mike Rose has noted, "Freshman composition originated in 1874 as a Harvard response to poor writing of *upper*classmen, spread rapidly, and became and remained the most consistently required course in the American curriculum" (342). If we aim collectively to give our students, as Robert Scholes urges, "the kind of knowledge and skill that will enable them to make sense of their worlds, to determine their interests, both individual and collective, to see through the manipulations of all sorts of texts in all sorts of media, and to express themselves in some appropriate manner" (15), doesn't it stand to reason that we would certainly, albeit indirectly, bring about a greater awareness in the public of social and political forces? We see this sort of awareness as necessary and an important first step. This awareness must be introduced, however, by teaching the production of written discourse and exploring what ramifications come from producing public writing, not from turning composition classrooms into classes about environmental politics. We offer some strategies for developing these sorts of classes in the next chapter.

But as we mentioned, we cannot count on collective efforts or larger changes in society. We can only hope to enable the students we come in contact with at least once a week to become more critical of the world around them and more efficient producers of writing. And we cannot expect to make changes in more than a few of their lives each semester. Once again, we need to avoid looking for sweeping changes and monumental efforts toward societal/ecological engagement. But what we must not overlook is the individual lives that we stand a very good chance of affecting in our classrooms and, in turn, how those lives will encounter others. As teachers, intellectuals "train critical readers and writers and prepare those who will move into positions of power and authority throughout society" (Merod, 39). We might look to our role in the classroom as a trickle-down effect. If we are able to empower just a few students each year to begin to "read" their own lives, develop their own opinions, and respond in meaningful ways to the environmental issues that affect our world, we will have affected a considerable number of individuals during the course of our careers. Moreover, we must coach our students to see that when they write, how they write effects their worlds. Rhetorical choices, invention strategies, writing processes are all part of a writer's ecology that influences their worlds. And these same students will soon become members, by and large, of the educated middle-class members who are very likely to influence the actions and opinions of others. And, while we cannot quantify how effective our efforts might be, it doesn't seem too optimistic to assume that they will be palpable.

While classroom assignments that superficially attempt to engage students in public debates—such as current events essays, letters to the editor, taking a stand papers, and the like—can "quickly seem absurdly decontextualized and formulaic in classrooms that are cut off from meaningful contact with the real public discourse of society" (Harris, 324), a thorough and concrete approach to public writing can substantially change the ways students read public texts and respond with their own. We provide, in the next chapter, a detailed account of pedagogical approaches for addressing the ecocomposition classroom. This kind of work helps students develop a critical self-awareness of their relation to the public world, and we would like to believe, as does Patricia Bizzell, that "such development leads in progressive directions" (par. 5).

Public Scholars

While many scholars might agree that classrooms are viable sites for encountering and producing public texts, fewer would agree that scholarship in composition studies is also a tenable space through which we can practice activism. Some argue that we are what we are only through clearly defined and rigidly structured modes of discourse, and to say things in ways unrelated to these particular modes is to risk losing our professional identities. But as scholars of discourse, our work can both be true to existing definitions of what it is we do and also move outside of our own discipline. For example, many compositionists have begun to concentrate on the internal rhetoric of discourse communities. All discourse communities have their own specialized means of producing and circulating knowledge, and these discourse communities use these specifically to represent and describe their particular outlook on the world. Analyzing these discourse communities "substantiates and extends the study of general rhetoric, which has been concerned with the specialized languages of discourse communities only as they enter into the shared space of public debate" (Killingsworth, 7). Scholars in composition who are interested in questions of public writing, rhetoric, and social/environmental activism might begin their scholarly work with studies of the various discourse communities that come into contact with each other through various public spaces. This places discourse specialists in a unique position within the academy both to "do what we do" and to extend our work to investigate intersections with other modes of discourse at the same time. Thus, we would reinforce our own disciplinary foundations through the study of other disciplines and discourse communities and how they circulate knowledge. Again, the contributions in *Green Culture* serve as examples of how such work might be carried out.

As the various discourse communities define themselves by their individual discourses, one result is that they often find themselves unable to communicate effectively with each other and often have difficulty entering the loosely agreed upon language of public debate. Specialists— whether they be academics, politicians, or members of other specialized discourse communities—are often unable to create strong communicative links with public groups, "links that would support a strong power base for reformative actions" (Killingsworth, 7). As discourse specialists, one of our goals should be to help create such links between discourse communities. This work might begin by devoting more attention to the issues that affect local communities and how they are resolved or perpetuated through discourse. As we've noted, ecocomposition intends to cultivate interdisciplinary investigations emerging from a number of seemingly disparate academic perspectives, including composition studies, environmental sciences, ecology, and others. By exploring the intersections of different discourse communities in public spaces, ecocompositionists might discover ways to build communicative links between different groups and individuals. That is, ecocompositionists must not only do the work that promotes ecological literacy, but also must do the work of discursive ecology, studying, for instance, the language used not only by ecologists but the language used by legislators to enact environmental laws.

If we begin to explore other discourse communities through our own scholarship, we might develop a greater awareness of how these various discourses come together in public spaces; how our own specialized type of discourse might benefit from exposure to the other types of discourse that we might encounter; and, ultimately, we might discover how we can best deliver our own specialized knowledge into contact with others in various public arenas. A critical examination of competing academic discourses is an important topic for eccompositionists. Understanding how academics interact among themselves through their scholarly journals and how they relate to different or competing discourses in public spaces might begin to teach us how we can best deliver our own messages to these groups. Such studies are deeply ecological, as they examine entities (academic disciplines) and the degree to which they compete for and share resources within an ecosystem (academia) and can provide an ecological study of discourse that leads to a better ecological understanding of the interactions between disciplinary organisms and the environment of the academy. By understanding the languages of other disciplines, ecocompositionists might gain the ability to speak to them in ways in which they are familiar (in ecocomposition, this might begin through conversations with our colleagues in ecology, as we suggest in the previous chapter). As a result, we might be able to bring together these seemingly dischordant perspectives to achieve some common goals, both within our universities and outside of them. Much of the intractability of modern social problems is due to the inability of concerned discourse communities to construct working relationship through language for the purpose of cooperative social action. Resolution and action are nearly impossible when different discourse communities are unable to communicate effectively. Careful studies of how different discourse communities generate and distribute knowledge might begin to give us the exigency we need to talk to a larger audience than "a few of our friends" (Fish, 1). Through our scholarship, we might begin to build bridges between our own work and that of other communities—all of which might begin to enable social/environmental change.

Public Workers

Our private efforts are necessary in bringing about public change through writing and physical work. Ecocomposition must recognize the material consequences of producing public writing and make use of the ability to write to effect change. Some of the most influential and important intellectuals involved in the study of discourse have recognized the value and importance of active work outside of the academy. Noam Chomsky, for example, has long been active in social and political debates. In an interview with Gary Olson and Lester Faigley, Chomsky comments on his work in linguistics and his political work outside of academe, stating that he has "two full-time professional careers, each of them quite demanding" (64). Chomsky even suggests that his social and political work is more culturally relevant than his work in linguistics, as he sees himself first "as a *human being*, and your time as a *human being* should be socially useful" (65). Paulo Freire was also noted for his work through both educational structures and through active political involvement. In a different interview with Olson, Freire asserted that "a progressive teacher, a progressive thinker, a progressive politician many times has his or her left foot inside the system, the structures, and the right foot out of it" (163). Likewise, in an interview shortly before his death, Michel Foucault emphasizes his urge to make social changes by "working inside the body of society" "participating in this enterprise without delegating the responsibility to any specialist" (160). Work outside the academy is vitally important; we cannot assume that attention to social and political debates inside of the academy is sufficient. While opportunities for social activism might not be as open to most of us as they have been for the intellectuals just mentioned, opportunities are nonetheless available. While some would argue that academics are too busy fulfilling the work requirements necessary to earn tenure and gain promotions or bogged down with excessive teaching and /or administrative duties, we might take an example from Chomsky, Freire, and others—all of whom found time for social activism in addition to active and productive careers.

Also, we bring our own lives into our scholarship and teaching. What we do as individuals in our private lives informs what we do as members of academia; productive intellectual work depends on our ability to import real issues into the work we do in our classrooms and that which enters conversations in our scholarly journals. Active work in public spaces translates directly into more dynamic and engaged scholarship and teaching.

Through these three interconnected means of engagement with public issues—through the classroom, through scholarship, and through our lives outside of academia—ecocompositionists might arrive at a new definition of activist intellectualism, one which takes into account the smaller roles and opportunities that are more readily available to us than the narrow definition of reaching vast audiences in short periods of time. We might begin to extend our definitions of activist intellectualism to accommodate the variety of opportunities we have to foster cooperative public connections. No one of these means is sufficient in and of itself, nor will we be able to bring about sweeping changes individually or immediately. But if we begin to refigure our roles as socially active intellectuals and understand that opportunities for activism are open to all of us, we can begin to have significant social and political effects. We need, as Ellen Cushman notes in *The Rhetorician as an Agent of Social Change,* "a deeper consideration of the civic purpose of our *positions* in the academy, of what we do with our knowledge, for whom, and by what means" (12). Activist intellectuals might be then, quite simply, members of academe who take steps to bring more voices, more discourse, and a greater degree of communication to public debates, and in turn bring about social change.

CHAPTER 5

Ecocomposition Pedagogy

༂

We teach our children one thing only, as we were taught: to wake up. We teach our children to look alive there, to join by words and activities the life of human culture on the planet's crust. —Annie Dillard, "Total Eclipse"

There is no such thing as an individual, only an individual-in-context, individuals as a component of place, defined by place. —Neil Evernden, "Beyond Ecology"

We need to revitalize our communities—including our educational communities, business communities, and political communities—so that the principles of ecology become manifest in them as principles of education, management, and politics. —Fritjof Capra, *The Web of Life*

By looking at writing ecologically we understand better how important writing is—and just how hard it is to teach. —Marilyn Cooper, "The Ecology of Writing"

Ecocomposition must be an active *praxis;* it must engage and involve students. It must encourage students beyond the classroom environment. Ecocomposition turns to Kenneth Bruffee's notion of the "conversation of mankind" and encourages students and teachers alike to participate in conversations about environment, place, and location, within and beyond the mapped places of classrooms. We would like to identify and discuss two primary branches of ecocomposition pedagogy. The first, and perhaps fastest growing in American colleges and universities, is that which lends toward ecological literacy, the kind Randall Roorda envisions in his CCCC talk. This pedagogy, essentially, teaches environmental awareness in the writing classroom. In ecocomposition, this pedagogy most directly evolves from the ecocritical teaching of environmental texts, frequently those texts which are labeled "nature writing" and from critically examining local and global environmental issues, much as other pedagogies have examined other cultural issues in order to spark classroom discussions and awareness. This form of pedagogy is best described as having evolved from the Freire-inspired concept of *critical consciousness,* which asks students to be critically aware of their world. This approach argues that the teacher is expendable and should, after working as a "change agent" in the classroom, dissolve his or her authority in order to allow students to emerge with a "critical consciousness" which allows them to reflect on issues and ideas that transpire in the political and social spheres they come in contact with in daily life. In ecocomposition, this form of pedagogy is manifested in the classroom as one in which environmental texts and issues are taught and analyzed much as other critical/political texts become the subject of the writing classroom. In this kind of pedagogy, ecological/environmental texts and issues become the foundations for students' writing. That is, in this form of pedagogy, environment and nature are the subjects about which students write, and developing a better understanding and a critical awareness of these subjects is the goal of the course.

The second form of ecocomposition pedagogy is what we will call "discursive ecology." Much like social ecology, which examines the relationships within and of societies, discursive ecology examines the relationships of various acts and forms of discourse. This branch of pedagogy asks students to see writing as an ecological process, to explore writing and writing processes as systems of interaction, economy, and interconectedness.[1] Like the ecological-literacy-based pedagogy, discursive ecology's pedagogy may indeed engage environmental texts, and may ask students to write *about* environmental/ecological issues, yet it emphasizes writing and discourse as the subject. In other words, the first version of ecocomposition pedagogy closely mirrors what seems to have become the conventional approach to teaching writing: the analysis of other texts or issues—which are often organized by either subject or genre—the teaching of written conventions, and the writing of papers which address the texts or issues (or similar texts and issues) first analyzed. For many, this form of ecocomposition will be the most practical to approach, and it certainly has been the most widely used eco-pedagogy thus far. We do not mean to suggest this variety of pedagogy to be an inferior or secondary pedagogy. In fact, we encourage teachers to begin their first ventures into ecocomposition with this sort of pedagogy. It is familiar and it asks students to engage ecological issues at the textual level. It urges students to think ecologically, it asks students to write about ecological/environmental subjects, and many good textbooks have been produced which support this form of pedagogy (see, for instance, textbooks by Anderson and Runciman, Arnold, Jenseth and Lotto, Ross, Verburg, Finch and Elder, and Halpern and Frank). We feel that the study of environmental and ecological writing is an important first step in that it can "plant the seeds" of eco-thinking and enables students to think in more holistic, systemic ways about their relationships with living beings and environments. Teaching ecological literacy enables students to understand the principles of organization of ecological communities (ecosystems) and apply those principles to their own lives and communities. However, what we must stress here again is that ecocomposition must emphasize the production of written discourse, that ecocomposition pedagogy must teach writing, not interpretation. We feel that the most progressive and dynamic forms of ecocomposition urge students to look at their own discursive acts as being inherently ecological. The study of nature writing may lend to helping individuals to think more ecologically, but we must also help students to see communication, writing, and the production of knowledge as ecological endeavors. Teaching nature writing as a literary genre is an endeavor for the literature classroom, not the composition classroom. That is to say, ecocomposition must be about more than simply bringing nature writing texts to the writing classroom; it must be about the act of producing writing. Asking students to read nature writing is not *doing* ecocomposition. As we hope we have established by now, writing is ecological, and understanding this allows us and our students to better grasp the intricacies of discourse and communication.

Environmental English Departments

Because environmental concerns have moved into English departments primarily through literary approaches, nearly all studies of pedagogical approaches to teaching environmental and ecological consciousness address the teaching of either "environmental literature" or "nature writing" with a few ventures into "environmental rhetoric." As Lawrence Buell (1995) has noted, "Environmental nonfiction, however, gets studied chiefly in expository writing programs and in 'special topics' courses offered as the humanities' title to environmental studies programs or to indulge a colleague's idiosyncrasies, rather than as bona fide additions to the literature curriculum" (8–9). The accurate point Buell indirectly makes is that many teachers of expository writing who do attempt to bring environmental issues to the classroom do so in ways that focus on literary studies rather than on the production of writing, and that the introduction of nature writing to English departments should, indeed, take place in literature courses—ones specifically designed to study nature writing. Certainly, ecocomposition has gained much from some of the endeavors to bring ecological and environmental thinking into the English curriculum, and its theoretical roots lie partly in the development of ecocriticism (see chapter 2). Given this relationship, little attention has been paid to the history of ecocomposition, as its history consists mainly of retroactively labeling work—such as that of Coe, Miller, or Cooper mentioned in earlier chapters—as actually being ecological approaches to thinking about writing or about composition studies, and creating a history of ecocomposition seems, at this time, premature. However, in terms of considering how ecocomposition pedagogies might evolve, it is crucial that we first look at how other environmental/ecological English studies pedagogies have evolved and what role writing has played in those pedagogies.

One of the most extensive assessments of the move of environmental concerns into English departments is Cheryll Glotfely's "Literary Studies in an Age of Environmental Crisis," which is the introduction to *The Ecocriticism Reader: Landmarks in Literary Ecology*. In this introduction, Glotfelty notes that Stephen Greenblatt and Giles Gunn's collection *Redrawing the Boundaries: The Transformation of English and American Literary Studies* proclaims that

> Literary studies in English are in a period of rapid and sometimes disorienting change. . . . Just as none of the critical approaches that antedate this period, from psychological and Marxist criticism to reader-response theory and cultural criticism, has remained stable, so none of the historical fields and sub fields that constitute English and American literary studies has been left untouched by revisionist energies. . . . [The essays in this volume] disclose some of the places where scholarship has responded to contemporary pressures. (quoted in Glotfely, xv)

Glotfelty immediately points out that none of the essays in Greenblatt and Gunn address ecological literature. Her critique is accurate and extremely telling of English studies' reluctance to turn not only to ecological literacy as a critical concern, but also to environmental literature as an important genre in English studies. "The absence of any sign of an environmental perspective in literary studies," she writes, "would seem to suggest that despite its 'revisionist energies,' scholarship remains *academic* in the sense of 'scholarly to the point of being unaware of the outside world'" (xv).

Glotfelty accurately identifies that "If your knowledge of the outside world were limited to what you could infer from the major publications of the literary profession, you would quickly discern that race, class, and gender were the hot topics of the late-twentieth century, but you would never suspect that the earth's life support systems were under stress. Indeed, you may not know that there was an earth at all" (xvi). Glotfelty continues to point out that in contrast, if one were to look at newspaper headlines, one would find a world riddled with environmental problems. She notes of particular interest that *Time Magazine* in 1989 named "The Endangered Earth" as person of the year. The discrepancy between literary studies' attention to environmental concerns and the outside world's concern is unconscionable. As Glotefelty says, "the claim that literary scholarship has responded to contemporary issues becomes difficult to defend" (xvi).

In 1996 when *The Ecocriticism Reader* was published, Glotfelty could easily make the claim that

> Until very recently there has been no sign that the institution of literary studies has even been aware of the environmental crisis. For instance, there have been no journals, no jargon, no jobs, no professional societies or discussion groups, and no conferences on literature and the environment. While related humanities disciplines like history, philosophy, law, sociology, and religion have been "greening" since 1979, literary studies have apparently remained untinted by environmental concerns. (xvi)

Fortunately, much of this has begun to slowly change—in part to Glotfelty's own efforts, and in part, according to Glotfelty's own assessment, because of the efforts of Frederick O. Waage and his MLA collection *Teaching Environmental Literature: Materials, Methods, Resources.* Following Waage, in 1989, Alicia Nitecki founded *The American Nature Writing Newsletter,* and in time a few universities around the country began to develop courses in environmental literature. Then, according to Glotfelty, "In 1990 the University of Nevada, Reno created the first academic position in Literature and the Environment" to which Glotfelty was hired. During this same time, more conferences began to recognize the growing study of literature and the environment and began to include more environmental literature sessions in their programs. Glotfelty notes the 1991 MLA special session "Ecocriticism: The Greening of Literary Studies" and the American Literature Association Symposium in 1992 entitled "American Nature Writing: New Contexts, New Approaches." And, in 1992 at the Western Literature Association meeting the Association for Study of Literature and the Environment (ASLE) was conceived. Until 1998, when

Randall Roorda initiated the conversations which would begin ASLE-CCCC Special Interest Group a year later in Atlanta, none of the discussions that led to the evolution of ASLE nor the movements which saw the introduction of environmental issues into English departments via literary studies addressed writing or the production of written discourse. While we are thrilled to be able to trace the genealogical roots of ecocomposition in this direction, it is both frightening to us that ecocomposition is merely a three-year-old inquiry if we take 1998 (the year both Roorda initiated formal conversations and Michael McDowell's piece was released—though Roorda's *JAC* piece had appeared at the end of 1997) as the birthday of ecocomposition and that prior to that date only limited attention to the relationship between environment and writing had been considered (Coe and Cooper stand as the primary exceptions, though Roorda, Cooper, and few others had led several discussions and panels at CCCC in the mid-1990s regarding nature writing and environmentalism). However, we don't wish to make such a claim without noting the ways in which the very few environmental literary scholars and teachers made efforts to incorporate writing into their pedagogies. A few examples must be noted.

In 1985, one of the first texts to provide teachers with methodological and theoretical approaches to introducing environmental writing into an English classroom was edited by Frederick O. Waage and published by MLA. *Teaching Environmental Literature: Materials, Methods, Resources* provides nineteen essays that "introduce college teachers to, and provide them models and ideas for, courses in environmental literature and, more broadly, courses that combine humanities and environmental studies disciplines, with literature and writing as major components" (viii). Though the collection emphasizes the teaching of environmental literature, Waage includes three selections which address writing pedagogies. The "major concern of the essays" collected in "Part Three: Nature Writing," according to Waage, is "the teaching of writing" (85). Yet, like many misunderstandings of both the composition course and of ecocomposition (though at the time of Waage's book, ecocomposition had not become a formal area of inquiry), the "teaching of writing" is often lost in favor of critical inquiry of the subject matter at hand. Waage, in fact follows his claim of the goals of this third section of the book by explaining that "Nature in many geographical areas of the United States is a 'text' that a whole class can 'read' together and is also one that can be 'read' by individuals as experience, not as words on a page" (85). Of course seeing nature as text is crucial; yet Waage's immediate reversion to reading reaffirms that "the teaching of writing" is often enacted through an emphasis on interpretation rather than production. That is to say, never does Waage ask if nature or environment can be written too, nor does he ask as to the impact the words on the page might have on nature.

Betsy Hilbert's essay in Waage's collection, "Teaching Nature Writing at a Community College," represents the sort of approach that, while perhaps useful for literature courses, should not be mistaken for an effective ecocomposition pedagogy. At first glance, the course described in this essay seems intriguing. Hilbert emphasizes reading a wide range of "nature writing" texts, including a variety of texts about local environmental issues and about the nature that surrounds her South Florida community college. Her reading list includes selections from Florida naturalist Archie Carr and Marjorie Stoneman Douglass, author of the classic *The Everglades: River of Grass.* Hilbert writes that "'Writing about nature' is the theme of this class not because it's my specialty or a rewarding area of literary studies and not because this study is central to some of the most crucial issues of our civilization—though all of those things are true—but because the study and practice of nature writing helps my students learn the things they need to know" (88). Her students are honors students enrolled in a freshman composition course which she has titled "Writing about Nature." The Florida state system has a Gordon Rule composition requirement specifying a required number of graded words submitted by each student; First-year composition courses at Florida colleges and universities are intended to emphasize the teaching of writing, and we assume (perhaps incorrectly) that Hilbert's class held a Gordon Rule requirement, though she does not mention it in her article.[2] The title "Writing about Nature" suggests that writing is a central focus of the course. Hilbert's class, despite its suggested intentions, does not teach composition, or at least her essay does not represent it as teaching composition. As we have said, teaching environmental issues or ecological literacy as the subject of a composition course can and often are vital parts of ecocomposition pedagogy, yet we wish to emphasize that such courses, when labeled composition courses, must use these subject areas not as the central components of the course, but as a vehicle through which students explore and produce *writing itself.* Granted, Hilbert's essay is also intended for a text called *Teaching Environmental Literature,* but her essay's position in the book (the first essay in the "Nature Writing" section) more than suggests that environmental literature can be brought into the composition classroom while maintaining the composition emphasis. In short, Hilbert seems to suggest that reading literary works can somehow substitute for instruction in writing itself. Her approach suggests that either (*a*) the "content" of the course—learning more about nature through the study of literary works—supercedes the actual production of the students' own texts, or (*b*) students learn to write more effectively primarily by modeling their work upon "literary" works. Of course, as compositionists, we feel that attention to the actual processes of writing is the most useful and constructive method of writing instruction; reading literary works does not sufficiently prepare student-writers. Hilbert's course description falls prey to a general failing of many composition courses which choose nature writing or other subject areas: the course redirects focus away from writing.

Early in her essay while describing the institutional setting and student population of her honors course "Writing about Nature," Hilbert explains that "the honors students read and write fairly competently, so that the range of our readings doesn't have to be limited" (88). This initial assessment of students' reading abilities is prevalent throughout the essay. In fact, this claim highlights a recurring problem with Hilberts' pedagogy: because students write "competently" the instructor can select a range of readings. That is to say, students' levels encourage Hilbert in regards to the texts that may be assigned, yet she does not mention the writing assignments that may be made. Similarly, Hilbert goes on to write:

> Why choose the theme of nature for a required composition course with these students? For one thing a solid dose of John Muir or Edward Abbey is just the thing for a class full of people whose overwhelming ambitions are to marry well and to become accountants. Nature writers have by and large been a bunch of mavericks—independent, tough-minded, much more inclined to explore Afghanistan or the creek behind the house than to settle into a regular job. Some may have passionately desired the good opinion of the world, but few have been willing to make the necessary compromises to get it. For another thing, nature writing exemplifies the writing skills that freshman composition is designed to teach, illustrating the challenges and problems faced by nonfiction writers. (89)

Again, the emphasis is focused on reading text. In addition, there is something suspicious about the agenda of disrupting goals of "marrying well and becoming accountants," particularly when Hilbert herself identifies that

> Many of our students have been out of school for several years or several decades, and they return to the classroom bringing a wealth of experience— sharp minds and rusty academic skills. We also get many students just out of high-school, most of them too young, too broke, or too badly prepared to go off to a four-year college. We see refugee and foreign students of all kinds and refugees from the ghettos of Miami, all of them convinced that

somewhere in the halls of our college they'll pick up a one-way ticket to the American dream, or at least a job in computer programming. (88)

And, yet, Hilbert does not mention how students' own writing might provide access to that dream or even to Hilbert's vision of a class full of mavericks. That is, never does Hilbert give credence to the rhetorical power of student writing, to writing as a productive activity. Instead, Hilbert contends that "a general competency to meet the freshman composition requirement might read: 'The student must prove that he or she is able to appraise critically and evaluate pieces of nonfiction prose with attention to content, organization and style'" (89).

It is interesting, too, if not more than telling, to note that Hilbert does claim that in this class "the constant emphasis is on learning to write," particularly when she immediately explains:

> because this class is a composition course meeting a general education requirement, not an elective class in environmental literature, we concentrate on methods of invention and rhetoric more than critical analysis, history of the literature, or environmental issues. It's sometimes hard to keep in mind that I'm teaching writing, not ecology or environmental studies, theories of nonfiction prose, attitudes toward wilderness in American literature, or any of the other five hundred absolutely essential topics that are far more interesting than anything a rhetorical textbook has to offer. (89–90)

The final editorial comment about rhetorical textbooks sums up not only Hilbert's approach to the teaching of writing, but reflects an often observed approach to incorporating environmental writing or any other subject-bound writing into a writing course: the writing takes a back seat. Hilbert, in fact, in the pages that follow this statement provides only a list of readings which students should read in order to model their own writing after established nature writers and then follows with pedagogical methods for addressing environmental issues in the classroom, despite her claim that "nature is the subject, not the content of the course" (92).

We don't mean to dismiss Hilbert's pedagogy, only to identify that ecocomposition asks for more from a composition course. Certainly, too, critiquing Hilbert's course as an ecocomposition course is difficult because of both chronology and the fact that she never claims the course to be an ecocomposition course. Yet, her claim that the course is a composition course deems that we address the oft-employed methods for emphasizing the readings and literary/textual interpretation over textual production, particularly when Waage has positioned this essay as one which emphasizes "the teaching of writing" and as the one selection which addresses a first-year composition course per se. Hilbert's course, it seems, mimics an unfortunately large number of first-year writing courses: one in which a particular literary genre or theme is explored and writing assignments are structured around that theme or genre. For ecocomposition, a course which emphasizes nature writing or local ecologies or environmental debates and offers writing assignments as a means to assess students on their ability to write about those subjects is a tacit approach to ecocomposition; it is a place to begin. However, ecocomposition must also ask students to consider the role of their own writing and rhetoric on those issues, texts, and environments. We encourage teachers to bring nature writing, environmental debates, and local concerns to classrooms, but we also encourage them to keep in mind that if the course is to be a composition course, it must first and foremost explore the production of writing with students. It must ask not only about the style and rhetorical choices of anthologized nature writers, but also about students' rhetorical choices, locations of students' writing, and the role of production in local environments. Hilbert's course does not seem to teach writing, but rather it teaches a genre of writing ripe for interpretation.

Like Hilbert's essay, Paul T. Bryant's is positioned as addressing the teaching of writing. In "Nature Writing: Connecting Experience with Tradition," Bryant explains that the literary genre of nature writing "can be related to the student's own writing, to the student's own experience, and to some of the major problems of contemporary life" (93). In order to make such connections in class, Bryant has devised a two-part course which addresses both the literary genre and the processes involved with writing nature writing: "the study of the literary tradition of nature writing is combined with writing assignments that require the students themselves to do nature writing from their own experience" (94). In order to accomplish this dual agenda, "the course has two lecture-discussion periods and a two-hour laboratory-workshop period each week. . . . The lecture periods are generally devoted to the study of the literary tradition and the laboratory periods to writing, peer criticism, and a few modest field trips that provide common experiences to be used in the writing" (94).

The first part of this divided class is structured like many literature courses: texts are read and discussed, background material and historical contexts are provided, cultural assumptions are questioned, and author's strategies are discussed. Bryant's course covers a wide range of texts from biblical through contemporary. Bryant writes that he spends only the first part of the course discussing older works as he is eager for students to spend most of the course examining contemporary nature writing as "the writers of this century can be used as models and sources of writing techniques for the writing the students are themselves doing in the laboratory. Consequently, from this point more attention is paid to matters of style, organization, rhetorical devices, and such, as well as to ideas and subject matter" (97). According to Bryant, analyzing contemporary nature writers "helps students become more sophisticated writers" (98). For many compositionists, this method of modeling is ripe with problems, but we shall leave the critique of modeling for now with the simple statement borrowed from Susan Miller: "reading is not writing."

For our purposes, it is Bryant's "laboratory-workshop period" in which we are interested. According to Bryant, "the two-hour laboratory session each week has become a workshop for writing, exchange of criticism, and brief trips into the field" (98). Bryant admits that there are advantages to this session, including the "obvious one of providing time for field trips," which we read as a testament to the attitude many teachers have toward the classroom time slotted for writing instruction: other "more fun" activities can supercede the actual teaching of writing. However, not to discount Bryant's pedagogy, he does describe a potentially useful beginning to initiating writing pedagogy in a nature writing course. He writes of the workshops:

> By asking students to work on a specific assignment at a given place and time, in a workshop setting that may include discussion of techniques and exchange of critical comments, we can emphasize that writing is a craft that involves the conscious use of describable techniques. We can demonstrate that revision is not only desirable but essential for even the most polished writers. And we create a situation in which students, in reacting to and offering helpful analysis of the writing of classmates, become far more perceptive in reviewing their own writing. (98)

Bryant structures assignments in this workshop to progress from personal expression, which emphasize the writer's role in nature, to a distancing of the writer from the work in order to provide the reader with "a new experience of nature through the medium of writing" (99). The transition of writing from the personal to the distanced seems reminiscent of Ken Macrorie's pedagogies. Of course, for our purposes, such a move is intriguing as it suggests a good deal about a writer's role in the creation of nature and the experience of nature for a reader. Yet, Bryant fails to examine the extensive role a writer has in a reader's experience of not only the nature described in a particular piece of writing, but also in all nature, all environment. In addition, Bryant neglects to ask his students to consider how the images and perceptions of nature about which they write became "theirs." That is, Bryant, while considering the processes of writing with his students strictly for the ends of the product, does not explore the relationship of writing to nature. But, Bryant is careful to note that "we move from writing as a private behavior to writing as a social act directed toward others" (99). Such a move from the expressivistic type of personal writing so often associated with environmental writing to a more socially oriented understanding of writing is critical. Bryant does well to guide students to achieve critical distance from their writing and "see it as something that is to be consciously crafted for a given purpose" (99). He explains that "when students see their work as an artifact that can be shaped and polished for a purpose, rather than merely a blurting out of their own ephemeral feelings, then it is possible to teach them style and rhetorical technique" (99). We applaud this direction in teaching nature writing; ecocomposition must address the social construction of nature and discourse, not simply the "feelings" or "self" of the author.

Of course, the largest problem with Bryant's pedagogy for our interest in ecocomposition is that it is primarily a literature pedagogy designed to ask students to produce literature, not explore the dynamics of producing writing—though it begins to move in that direction. It does not ask students to examine the ecologies of those processes, or even to consider the ramifications of their writing on nature. Again, we do not mean to set Bryant (or Hilbert) up as "straw men," since their intent was to describe the role of writing in courses specifically designed to teach nature writing both as literature and as the culmination of a specific kind of writing pedagogy. What we offer here by way of example is how ecology made initial moves into English studies.

Waage includes one final essay in the "Nature Writing" section of his collection. Margaret McFadden's "'The I in Nature': Nature Writing as Self-Discovery" uses nature writing as a vehicle through which students explore "the problem of the thinking and writing self, the autobiography of the self and its relation to nature" (103). McFadden wants her students "to get used to writing about themselves and their relationship to their environments" (103). In order to promote this goal, McFadden dedicates a large portion of the course to what she calls "I in Nature journals" in which students write about themselves and their observations of nature and the course readings (103). Unfortunately, McFadden's course is strictly based on students' exploration of the readings; writing is required, but not taught. The course described here is more of a nature-writer-as-genre course not even a how-to-write-nature-writing course as was Bryant's. Even more unfortunate is the fact that such a course is considered a writing course and that as such it concentrates on self-expression—an approach that is akin to expressivist notions of composition advocated by Peter Elbow, Donald Murray, Ken Macrorie and others in the 1970s—rather than exploration of the discursive construct of the concept of "self" and what role such construction plays in the expression of nature in text.

One of the things that we would like ecocomposition to move away from is the neoromantic notion of the solitary author (usually a male) who communes with Nature and discovers some truth about either himself or the world as it "really is." The problem with such notions is that they all too often portray the author or protagonist as the authority, the objective scientist, or the sage and do not acknowledge his dependence on and imbrication in a series of systems; this author-as-authority approach is reminiscent of Donna Haraway's notion of the "god-trick." In order for ecocomposition to become a truly ecological endeavor—in the sense that we have described in chapter 3—it must recognize that knowledge, truth, reality, even identity, emerge as a result of a complex array of influences, both human and non-human. As Christian has argued elsewhere, "Our identities—and all knowledge that we have of our selves and of the world—are shaped and enriched by our contact with others in specific settings, and the degree to which we maneuver around, beside, and in accord with others has much to do with who we are." This is not to suggest that communing with nature is anti-ecological; only that viewing oneself as solitary or isolated while in nature is anti-ecological. In fact, in chapter 6, we will argue that the sort of emotional-ethical appeals so common in nature writing often advance ecological thinking—as long as they do not view the solitary individual as separate from ecological systems.

Ecocomposition pedagogy must consider the role of discourse in the construction of knowledge about the world—including what we know (or think we know) about nature. It must theorize nature as a discursive construct while maintaining a recognition of the biological, chemical, systems of "nature" and our role as living organism in those systems (though we recognize those systems only through discourse). That is, advocating an ecologically sound pedagogy (or research)—one which emphasizes place, environment, or even "nature"—is not a call to neoromantic, expressivistic connections between self and the world. Rather, it is a call to extend the view of discursive construction beyond the individual, the social, and the cultural. It is a call to see that environment both contributes to these constructions and is a part of them. By not being conscious of surrounding environments (natural, constructed, imagined, etc.), we limit our view of the world, of our influence on and in that world. We deny our roles as living organisms. We deny how we construct that world, how we know that

world. We certainly have individual connections, feelings, and spiritual-
ities about specific places, environments, and natures; we do not wish to
sound like unfeeling stoics. But we recognize that such connections are
discursive, and such an awaking (like those to which Annie Dillard alludes
in our epigraph) enhances those connections by providing a more accurate
understanding of those places. Ecocomposition pedagogy must not aban-
don our emotional ties to various natures, but it must acknowledge the
ideological, discursive sites from where those connections come.

In the years following Waage's MLA collection—during which composition studies became a much more vocal and prevalent entity in the American academy—more scholars of writing began to develop more ecologically sound pedagogies, although few incorporated the discursive ecology
pedagogy during this time. In fact, the early 1990s saw an explosion of
interest in the inclusion of environmentalist approaches in a wide range of
academic disciplines. One book that influenced scholars across disciplines
to incorporate ecological and environmental focuses in their courses was
Jonathan Collett and Stephen Karakashian's (1996) *Greening the College
Curriculum: A Guide to Environmental Teaching in the Liberal Arts*. In their
collection, Collett and Karakashian explain that they often envision college teachers who teach "Intro" courses for majors pondering how they
might incorporate environmental concerns into their courses. They accurately imagine these teachers to be most concerned with using class time to
convey the important subject material for that introductory course, no
matter the subject. After all, they, note, wasn't this the task for which the
teacher was hired? "How could she include anything on the environment?
And, isn't it the province of the new Environmental Studies Program" to
address such concerns (1)? Collett and Karakashian note that this perception mirrors attitudes toward African-American Studies and Women's
Studies held by many in the academy during the 1970s and that it has
"taken until the 1990s for the curriculum, textbooks, and individual faculty to catch up and begin to represent issues of race and gender in a wide
range of liberal arts courses, although still not without controversy" (1). In
turn, what Collett and Karakashian contend is that "we cannot wait 20
years to make it possible for the majority of students in the liberal arts to
confront the challenges of an environmentally sustainable future" (1).

What Collett and Karakashian provide in their collection is a guide for
liberal arts teachers for introducing environmental materials into their curriculum. They argue that their collection will do the following:

> (1) It will provide her [the instructor] with a rationale for including material
> on the environment in the teaching of the basic concepts of her discipline.

(2) It will show her how to construct a unit of a full course at the introductory level that is a basic course in the discipline yet makes use of environmental subjects. She may be ready to propose an upper-division course in her department, perhaps cross-listed in Environmental Studies, and *Greening the College Curriculum* gives her sample course plans, with a wealth of ideas about bringing the subject matter to life.

(3) It serves as a compendium of annotated resources, both print and non-print, for materials in her own related disciplines. (2).

In the pages that follow, Collett and Karakashian's collection contains twelve approaches to introducing environmental materials into the liberal arts curriculum. The opening essay, "Reinventing Higher Education," by David W. Orr, is perhaps the most useful piece, and for those interested in how and why American education should be rethought in terms of environmental awareness, Orr has much to offer. However, for our purposes in looking at pedagogical development, it is the other twelve chapters to which we will momentarily turn our attention. Collett and Karakashian provide chapters which address anthropology, biology, economics, geography, history, literature, media and journalism, philosophy, political science, and religion. Notice, not only the absence of writing, but of English in general. "Literature" becomes the English studies representative in this collection. In our experience, very few "Introductory" courses in English departments are intro literature courses anymore. In fact, initially seeing the title of chapter 7 listed as "Literature," we immediately turned back to the copyright page to see just how outdated this book was. To our surprise the book was published in 1996. Long before then, first-year courses in English had turned their attention toward composition; long since have English departments supplanted literature departments. Yet, literature was to be the connection to environmental studies in English for Collett and Karakashian.

In the chapter on literature, author Vernon Owen Grumbling writes that "as the enduring record of creative imagination, literature potentially encompasses all subjects of human thought and experience" (151). Most of Grumbling's pedagogical approach involves exploring a variety of literary and interdisciplinary texts which address environmental concerns. Grumbling does require students to keep journals, for which he provides writing prompts, but no writing pedagogy appears. The course is as much a traditional literature course as one could imagine. Our concern here isn't with Grumbling or his pedagogies, but rather with the dismissal of writing pedagogy throughout not only Collett and Karakashian's book, but throughout the overall environmental move into English studies (with the exceptions we have noted of Roorda, McDowell, and a few others) until very recently. What concerns us most is not that courses often called "Nature Writing" have made their way into more and more university curriculum, but that they are actually courses which address the genre of nature writing as literary critique, not as issues of textual production, yet claim to be writing or composition courses. We encourage the development of more "Nature and Writing" courses, but we urge those designing and teaching such courses as composition courses to foreground the *production* of texts rather than their consumption.

Ecological Education

Like Glotfelty's concern that American literary studies has not turned its attention to the environmental crisis in the same productive ways as it has toward race, gender, and culture, education theorist and reformer C. A. Bowers asks American educators—in both public schools and universities—to rethink their approach to educational reform.

> if the thinking that guides educational reform does not take account of how the cultural beliefs and practices passed on through schooling relate to the deepening ecological crisis, then these efforts may actually strengthen the cultural orientation that is undermining the sustaining capacities of natural systems on which all life depends. How our cultural beliefs contribute to the accelerated degradation of the environment . . . is the most fundamental challenge we face. All other social and educational reforms must be assessed in terms of whether they mitigate or exacerbate the ecological crisis. (*Education,* 1)

This theme runs through several of Bowers' books as he offers various critiques and approaches for education reform directed toward a more ecologically sound approach to education. Bowers offers that education, an institution whose primary function is to reinscribe cultural knowledge, works within a mindset that assumes that progress and development are necessary, normal, and needed. He questions the value of teaching a culture of consumption while disavowing other cultural literacies of ecological sustainability. Bowers' work is crucial to educational reform. We concur with Bowers' desire to develop a curriculum that questions cultural assumptions about ecological consumption and teach nothing of sustainability. We also agree that such education must occur at the earliest stages of learning, for as Bowers writes, "how children experience meaning and choices, interpret the nature of relationships, and make moral judgements reflect the deep and generally unconscious influence of culture" (*Educating*, 75). Ecocomposition gleans this notion that education should become more responsible for the inscription of ecological values in students. Hence, as we have said, ecocomposition takes as one of its primary goals the desire to encourage students to consider the relationships between (written) language and the earth's systems in which they must survive. And while we agree with Bowers that asking students to participate in a culture of sustainability must occur at all moments of education, it is our purpose to explore how that might be approached through the teaching of writing.

It would seem obvious that the first step in promoting a pedagogy which asks students to consider the world in which they live and the effects of writing in and on those systems would be to make "the world" the subject of the class. That is, just as many progressive or (so-called) radical pedagogies have asked students to think, read, and write about culture, about contact zones, about conflict, and about difference, ecopedagogies must begin by asking students to write and think about subjects of ecological significance. In fact, we would argue that an ecologically sound pedagogy might very well encompass those very same progressivist topics and extend beyond them to address their implication in the living systems of which they are a part. Fortunately, several composition scholars have begun to address these very issues from a variety of theoretical perspectives, providing new ways to both think about and teach writing. Three books in particular address environmental/ecological pedagogies within the classroom: Randall Roorda's *Dramas of Solitude: Narratives of Retreat in American Nature Writing,* Derek Owen's *Composition and Sustainability: Teaching for a Threatened Generation,* and our own collection *Ecocomposition: Theoretical and Pedagogical Approaches.*[3]

Randall Roorda begins his 1997 *JAC* article "Nature/Writing: Literature, Ecology and Composition" by pointing out that "I am a peripheral figure: a compositionist specializing in nature writing" (401). Roorda goes on to explore the role of nature writing in composition studies, examining "those two loaded terms, 'nature' and 'writing'" (401). Part of Roorda's linking between nature and writing and nature writing is composition's long-held interest in literary nonfiction. For Roorda, nature writing is "the sort of writing—the essay especially, personal or informational—that a great number of us teachers exhort our students to produce, and so it behooves us to comprehend its operations" (402). Integrating the essay, traditional literature, and nature writing into composition studies, Roorda posits, "might even be thought to promote the transcurricular reconfigurations of composition, literature, and other disciplines" (402). According to Roorda, through such writing, "composition enters a claim on the 'literary' without giving itself over to literature or resigning itself to separate but equal status" (402). Roorda justifies this intersection between the "literary" and the disciplinary realm of composition claiming that "the intricacy and dash of literary analysis may be mistaken for pedagogical effect, the upshot tending to reinforce prevailing splits between the author's, the critic's, the teacher's, and the student's production of meaning—splits upon which composition's separate status is largely founded and that attention to nonfiction texts might be expected to address" (402). In the pages that follow, Roorda goes to great lengths to not only link nature writing to the study of writing and to the writing classroom, but also to consider as well the role of particular literary genres in the composition classroom. The year following the *JAC* article, Roorda would ask compositionists to again consider the role of nature writing in composition in his important book *Dramas of Solitude*.

Dramas of Solitude is not, at an initial glance, a book of either composition theory or composition pedagogy per se. It is a book about American Nature Writing; it is a book which would seem more at home on the shelf of literary critics, not on the shelves of composition scholars and teachers. But as Roorda notes early in the book, as nature writing becomes a more predominant area of literary studies, it also begins to show signs of having the potential to initiate large scale change in, as he quotes famed nature writer Barry Lopez, "a reorganization of American political thought" (1–2). Nature writing's interdisciplinary, its political charge, and its potential for initiating change, Roorda claims make "the genre an object of interest to rhetoric and composition as well as literary studies within English studies proper" (2). And certainly Roorda is correct. Like Bowers and others before him, Roorda identifies the crucial need to explore nature writing— that is, writing about environmental concerns—in order to promote a greater consciousness of environmental crises. And, we would again like to note the importance of both this book and Roorda's other efforts to bring such concerns to the writing classroom.

Like Bowers, Roorda asks readers to consider the roles of literacy and education in our approaches to understanding nature. His claims often mirror Bowers':

> If education is charged in part with preparing its wards for the workplace of the future, it might entertain the notion that any sustainable cultures to come must depend less on the "virtual" and more on the actual work of the light-powered body in the local milieu. It will not perforce turn scholars into farmers, yet it might devise heuristic devices to cultivate attitudes conducive

to this shift—an "environmental education" broadly conceived. Education in literacy I would think to be central to this change, with the narrative of retreat an exhibit in its perils and potentialities. (216–217)

For Roorda, the narrative of solitude—what he terms "retreat narratives"—exemplify specifically the kinds of literary works which provide student writers with access to ecological literacies as well as models for texts they might produce on their own. Roorda examines a variety of student writings in conjunction with critical readings of major figures in the nature writing genre—Wendell Berry, John C. Van Dyke, and Henry David Thoreau—in order to, first, help define the genre of nature writing and, second, according to Christopher J. Keller, to make "retreat narratives important to rhetoric and composition studies by questioning how student literacy and identity are formed; how writing, or composing, differs in solitude and society; and finally, how rhetoric and composition as a discipline might be restructured by new examinations, definitions, and uses of these texts" (511).

Classifying *Dramas of Solitude*'s role in composition—even within ecocomposition—is slightly problematic. At first glance, Roorda seems to be arguing for a greater inclusion of reading literary (nature writing) texts in the composition classroom. In many ways, for us, this seems a counterproductive maneuver, one which emphasizes the literary tradition of interpreting texts over the concern with the production of texts. However, Roorda's agenda in doing so is one linked not with improved literary analysis, or even for furthering the establishment of nature writing within the more often taught/studied literary genres, but instead with the goals of increasing ecological literacy while at the same time raising students' awareness of the rhetorical situations that are involved in "retreat narratives." His approach to nature writing—to ecocomposition, if we can call it that, for it is not a term he uses in *Dramas of Solitude*—is one that falls within the spectrum of cultural studies and composition studies. As Keller notes, "Cultural studies methodologies may be prominent in composition studies, but *Dramas of Solitude* pushes these boundaries further by posing questions of ecological literacy, questions that ask how we shape and are shaped by place(s), in and out of the literary. Such questions are crucial in forming more comprehensive notions of student identities" (512). As we mentioned in chapter 2, these links with cultural studies are important, and questions of the influence of place and environment on identity must be foregrounded in both cultural studies and composition. However, despite composition's rather vocal concern with identity formation and identity politics in recent years and our own call to examine the role of place in those studies, ecocomposition must move to explore the role of writing and place in inquiries about identity—as well as the role of place and identity in writing. *Dramas of Solitude* pushes ecocomposition in the direction of those questions when it asks us to consider how writers write in nature.

For the most part, however, what Roorda is leading to is the first branch of ecocomposition pedagogy: ecological literacy, which asks students both to consider how others approach nature in their writing and then to experience and write about nature, to place themselves in nature and consider why and how they are there. Ecocomposition embraces this methodology; after all, ecocomposition must be an active, participatory pedagogy. It must ask students to not only consider their roles in systems, both natural and constructed, but to place themselves in those environments and look at the material consequences of being there, of writing there. While Roorda's work is useful in that it provides a fuller understanding of ecological literacy and the role of nature writing in composition classrooms, Derek Owens' *Composition and Sustainability: Teaching for a Threatened Generation* extends the discussion beyond the composition classroom in an effort to explore ecological thinking in a broader pedagogical location—the entire university curriculum.

Owens stipulates that *Survival and Sustainability in the New Curriculum* is a book that argues for "reconceptualizing composition studies as a workplace wherein educators might imagine a new kind of curriculum through the metaphor of sustainability" (3). It is also an appeal to educators not directly affiliated with composition studies to entertain the prospect of composing their disciplinary objectives via what he refers to as a "reconstructive consciousness" (3). In doing so, Owens calls into question the "limitations of disciplinary thinking and the academy's compartmentalization of knowledge-making" (3). It is important to note that for Owens, being a compositionist entails more than just teaching writing. He argues:

> People who work in composition studies—who teach writing courses, fashion writing and writing across the curriculum programs—are certainly an intended audience. But so are members of an amorphous group comprised of students and teachers interested and working in environmental studies, educational philosophy, ecology, economics, art, architecture, labor studies, cultural studies, futurist studies, and service learning. The readers I have in mind are educators who, regardless of their disciplines, share what might be considered compositionist impulses. (1–2)

According to Owens, compositionists are interested in the art of composing, of putting things together, of "combining, arranging, mixing, and assembling: constructing with words or images or sounds (or all three) in virtual or physical space" (2). Certainly, we would agree that in order for composition to truly become ecological, it must reach across disciplinary boundaries and must involve students and teachers from a variety of disciplinary backgrounds. We also agree that the acts of combining, arranging, and mixing share many parallels in both the production of discourse and ecological thinking.

In the course of his text, Owens, like Bowers, calls for curricular reform that takes into account a more holistic vision of education, not allowing composition, or any other discipline, to stand as an entity separate from the rest of the academy. Owens' claim is that specialization—like that required in many graduate programs which require a singularly-focused dissertation as the culminating measure—creates systems which resist the relational and instead emphasize departmentalization, separation, and disciplinarity. Such separation, Owens contends, denies the relationships between various knowledge-making entities, and in turn, resists an ecological view of the world. Instead, Owens argues for what he calls "Bee Vision" of the curriculum. As he explains it, Bee Vision adopts and applies mosaic theory to curricular thinking:

> mosaic theory attempts to describe how arthropods with compound eyes see. Bees see more than one way. The bee eye discriminates differently depending upon its on-line (in flight) location and in accordance with its aim. . . . In some cases a 2D-pattern vision is used as the bee hones in on a target, distinguishes it by edge patterns. At other times the bee measures the range of landmarks in 3D. The mechanisms by which insects actively see in flight by altering their head movements in response to their environments, automatically reconfiguring their vision depending upon their position, are being studied so that they can be written into software for robot vision. (11)

He continues that such thinking should influence curricular design as "conceptualizing curriculum should be an exercise in cross-pollination. It has become necessary to think like bees see, missing neither the forest nor the trees" (11).

Turning to David Orr's ecological critique of scholarly and pedagogical work, Owens cites Orr's now famous and important heuristic that all academics should consider: "Where does your field of knowledge fit in the larger landscape of learning? Why is your particular expertise important?

For what and for whom is it important? What are its wider ecological implications and how do these affect the long-term prospect? Explain the ethical, social, and political implications of your scholarship" (quoted in Owens, 13). Owens argues that compositionists are in the best positions to answer these questions in ways that will help develop ecologically sound curricula. He contends that "it is exactly the permeability affiliated with composition that makes it an appropriate field for developing approaches to imagining sustainable curricula" (13). He posits further that despite the degree to which writing courses are generally embedded in English departments, they are mostly "extra-disciplinary" (13). Owens goes to great length to explain why composition is ripe for becoming the initiating discipline in the academy for developing sustainable, ecological, interdisciplinary curricula. We quote part of Owen's explanation at length here to help situate our critique and our encouragement of his work, but note the difficulty his agenda has for composition:

> We might strive to envision composition, and in fact all education, as environmental studies—not an offshoot of ecology but literally the study of one's immediate, everyday local environs (city blocks, strip mall parking lots, back yards, office cubicles, crowded highways) so that students can begin to articulate what's wrong with these places, explore how their identities have to no small degree been composed by such places (and vice versa), and creatively imagine ways of transforming them. Because of their extensive contact with first year college students, composition courses can serve as filters through which undergraduates might come to regard the objectives associated with their chosen majors from a growing sense of sustainable awareness. Because composition courses are so open-ended, and because composition teachers have more autonomy than most to bring a cross-disciplinary range of subject matter into the course, writing faculty can design classroom atmospheres that might get students to begin a process of critiquing their identities, needs, and desires in relation to their consumer culture in the formal limits of their local environments. Composition studies can be reframed as a disciplinary vehicle for promoting sustainable pedagogical philosophies. Composition as a service discipline— where "service" means helping students see the value in relating writing, critical thinking, and discipline-specific concerns to the conditions on their local living spaces. And because composition provides a forum for students to testify about the problems in their immediate neighborhoods and workplaces, and their hopes and fears for the future, composition is also a field for servicing the academy: making faculty and administrators continually aware of what our students live through on a daily basis, and forcing us to

confront the degree to which such realities do or do not play a role within curricular design. (14–15)

In essence, it seems that what Owens is calling for is a reshaping of composition away from a writing-based discipline to a discipline that uses writing as a vehicle through which sustainability is promoted. Granted, sustainability must be promoted not only in composition, but in all academic fields and all daily living. However, when Owens foresees the critique that such a move will be seen as shifting the focus away from writing production toward political agendas, his response is all too familiar; it is one often used as simple retort to the same critique leveled against cultural studies or other social politics in the composition classroom. It is an easy response to say that cultural studies, gender studies, or even sustainability studies will promote critical thinking and help students become better writers and better people. However, this maneuver is once again a move away from writing, a move away from seeing writing as the subject that should be studied in a writing classroom. While the secondary subjects that appear in composition classrooms—the "open-ended" more autonomous "cross-disciplinary range of subject matter" to which Owens refers—may provide these kinds of opportunities in some ways, the primary disciplinary focus of composition classrooms must be the production of writing.

It is also interesting that Owens suggests composition as the place to initiate a larger campus-wide move toward sustainability. While we agree that composition classrooms must address not only sustainability, but also the ecologies of writing and the ecologies of writing in larger environments and the affects upon those environments, we question the position Owens gives to composition. If we briefly examine the ecology of composition studies within not only English departments but also within larger institutional environments, we have to acknowledge that composition sits near the bottom of the food chain—an uncomfortable position to occupy, and a difficult place from which to initiate anything. According to Susan Miller, composition has, for decades, "conveniently and precisely contained within English the negative, nonserious connotations that the entire field might otherwise have had to combat" ("Feminization," 45). Composition has been marginalized, and it is seen as a discipline that is in the service of English departments and universities. We do feel that composition has the potential to initiate sustainable, interdisciplinary moves, but we feel that it is necessary to recognize the political frameworks that encumber composition's current position within the academic ecosystem.

Having said that, however, we do not wish to promote a cynical view of either Owens' important contribution to the development of ecocomposition or to what ecocompositionists can do. As we mentioned in the previous chapter, being an activist intellectual requires that classrooms become sites for activism, for beginning our work as ecocompositionists. However, in doing so, we must acknowledge first and foremost our roles as compositionists. Our classrooms cannot be just about the politics of environmental crisis, they must be about writing.

Like others who have begun to develop ecocomposition classrooms that focus on reading nature writing and discussing ecological awareness, Terrell Dixon, in his recent essay, "Inculcating Wildness: Ecocomposition, Nature Writing, and the Regreening of the American Suburb," introduces readers to a class designed specifically as a class in ecocomposition. As we noted in chapter 1, Dixon defines ecocomposition classrooms as "classes that emphasize reading and writing about nature and the environment" (77). As is now evident, we see this definition of ecocomposition as only a part of what ecocomposition is/should be. Ecocomposition must be primarily about the production of written discourse, not only about nature writing. The agenda of Dixon's course was to introduce students to a variety of canonized nature writers "as nature writers from whom students could learn both ecological values and the techniques of good writing" (77). These are certainly noble goals, and we urge students and teachers to concentrate on developing ecological values. However, the course described by Dixon, as it is represented in his essay, is not ecocomposition.

Dixon, by self proclamation, is an ecocritic, and by our accounts, he is a fabulous ecocritic. His course design defined in "Inculcating Wildness" reflects his commitment to ecocriticism, nature writing, and ecological awareness. He specifically identifies that his course (co-taught with another senior faculty member and six graduate students and enrolled by "180 first-semester freshmen") was an "experiment in teaching environmental literature" (77). In the pages that follow, Dixon offers a wonderful description of the intersections of natural and urban environments (and the problems of those intersections) surrounding Houston and the University of Houston Central Campus where the course was taught, particularly Brazos Bend and Greatwood. This description is an example of masterful nature writing in and of itself. He then offers ecocritical readings of the works of three authors whose writing addresses local environments and whose works were read by students in the class: Rick Bass, John Hanson Mitchell, and Robert Michael Pyle. Following Dixon's lengthy ecocritical situating of these authors' works within the context of both the course and the issues of Houston's urban/environmental contact zones and what these authors can teach about ecological awareness, Dixon returns to the class setting itself for a brief two pages. In the concluding section, Dixon explains that under administrative pressure, he and his co-teachers felt the "usual pressure to succeed" coupled with a sense that the large class—uninterested in the readings, by his claim—was drifting away (87).

Dixon and his co-teachers sent their "students out on campus to write about where the trees were, about what kinds of trees they were, how healthy they were, and why" (87). He continues: "We encouraged them to write about campus fountains, squirrels, and, yes, even, starlings" (87). From these writings, the class turned to looking at the city. "All of this," claims Dixon "helped move the notion of environment from abstraction to a tangible concern" (87). Such strategies mirror those of others whose classroom emphasis is on using nature writers to promote an ecological awareness among students. As Dixon tells it, this class became an important learning experience not only for students—many of whom Dixon encountered again in later classes—but also for himself and the other teachers as well.

> We learned something about trying to teach environmental literature in our own urban place, about how to make those absolutely crucial pedagogical and personal connections between a threatened wilderness that cries out for preservation and an urban park that needs native plant restoration, wild butterflies, and some wild space. We learned what urban nature writers like Rick Bass, John Hanson Mitchell, and Robert Michael Pyle seek to teach us: In contemporary America, we must connect the remote wilderness preservation so important to us with the cities where most of us now live. Because a good part of our urban commuter student population does not have a family tradition of summer backpacking trips or expeditions, ecocomposition and environmental literature need to emphasize that its concerns with wilderness encompass the city as well as the country. Brazos Bend and Greatwood as well as the Arctic National Wildlife Refuge and the Brooks Range. (88)

We could not agree more with his urge to connect wilderness preservation with urban life; however, we are baffled by Dixon's use of the term *ecomposition*. Throughout this essay, Dixon focuses solely on ecocriticism, environmental awareness, and nature writing as text for interpretation. Other than one brief mention of sending students to write about trees, Dixon does not address ecocomposition, composition, or writing in any way. Ecocomposition must emphasize the production of written discourse; it must emphasize rhetoric; it must emphasize the teaching of writing. Ecocomposition is not ecocriticism. Doing ecocriticism in a class of "first-semester freshmen"—a course traditionally aligned with composition—does not turn ecocriticism into ecocomposition. We do not mean to dismiss in total Dixon's article, as it is a fine piece of scholarship (barring the ecocomposition reference). But, while we think Dixon's article is both interesting and well-written, we are concerned with the prominence he attributes to ecocomposition by placing it as the primary term on the right side of his colonated title and then not writing about ecocomposition in ways other than to offhandedly invoke the term. We are wary of the appropriation and use of the term as an extension of classes that are an "experiment in teaching environmental literature" and then only use nature writers as models for students (77). Of course, we encourage Dixon and others to integrate nature writing, environmental awareness, local issues, and the likes into English classrooms, and to produce scholarship about those experiences, but we must be cautious not to cubby hole ecocomposition as writing "about where the trees were" (87). Doing so would be disastrous.

To be fair, though, because so much ecopedagogy initially came to English classrooms through ecocriticism, it seems natural that pedagogies that stress interpretation over production have evolved in similar fashion. Certainly our own moves into ecocomposition reflect the move from ecocritical approaches in a writing classroom to ecocomposition's focus on production. Sid's 1999 *Composition Studies* article about a course in environmental rhetoric and expository writing, for instance, plainly explains that his course "emphasized that environment is as important an issue in cultural studies as are other critical categories" and that the students were asked to engage "close readings and analyses of newspaper articles, essays, publicity material (particularly that of the Florida State Park system), and laws" (80, 83). Though Sid's class was grounded in students' production of public writing, his article makes clear that there was also a dual goal of "examining written (published) texts which contribute to some of the more visible environmental debates in Florida" (87). That is to say, ecocomposition's focus on production instead of interpretation has been much needed, but it has been precipitated by the prevalence of literary criticism's—particularly ecocriticism's—influence in English departments and the misconception that ecocomposition is about reading environmental texts and nature writing.

Ecological Literacy Approach

What we would like to offer now is a pedagogical approach—and a number of possible assignments—gleaned from an amalgam of ecocomposition courses we have taught at the University of Florida, University of Tampa, and University of Hawaii at Hilo and taking into consideration our vision of ecological literacy, public intellectualism, and ecocomposition. As we have shown, the ecological literacy approach to ecocomposition has been the predominant manner in which concepts of ecology have moved into composition studies. Most broadly, this approach to teaching ecocomposition regards the students' awareness of the importance of "place" as a central goal of the course. For the most part, these pedagogies have stressed a greater awareness of the "natural" world, of the environmental crisis, of the role of human beings in the destruction of environments, and in developing sustainable ways in which to continue to live on the planet. Such courses, it is argued, enable students to think more critically about the world they live in, and critical thinking is a necessary precursor to effective writing. We agree that ecocomposition courses focusing on these topics could potentially cultivate critical thinking. However, these courses need not focus exclusively on "natural" places, nor should they present the teacher's environmental-activist approach as the only perspective. As we've suggested, ecocomposition must strive to recognize the importance of all locations—not just natural ones—on individuals' lives, identities, and ways of thinking. In other words, ecocomposition courses could potentially, and occasionally do, focus on such topics as the ways that city-dwellers develop certain patterns of behavior or how Internet chat rooms allow individuals to come together in "locations" that best suit their needs. People develop through their interactions in a variety of locations, and while there is much that we can teach and learn through the study of the natural world, this topic is by no means the only "place" worthy of examination. Compositionists in downtown New York, Los Angeles, or other metropolitan areas are not unable to teach ecocomposition just because they cannot take students to pristine forests, rivers, deserts, mountains, or beaches; the places their students come from and exist in are equally deserving of study. Ecocomposition pedagogies must expand to encompass all sorts of places and their importance to people, groups, communities, and cultures, not just the stereotypical "natural" places.

In addition, ecocomposition pedagogies must seek to expose students to positions and beliefs that often run counter to the teacher's "liberated" perspective. It is easy enough to force-feed students the sort of texts that you agree with, and it is equally easy to train students to say what you'd like to hear. However, if critical awareness is a real goal, the best and perhaps only way to help students to become more critically aware of the world and their place in it is to expose them to the profusion of attitudes, ideas, ideologies, and perspectives that pertain to place or location. Interestingly, exploring "natural" environmental issues is one of the most effective ways of inquiring into that very profusion. Debates about environmental issues occur in a wide variety of institutional and cultural locations, and the actors in these debates range from the ultra-conservative to the extreme liberal. Only by exposing students to this multitude of perspectives can we hope to enable them to develop mature and enduring positions of their own.

In addition, ecological literacy pedagogies and assignments should be designed with larger, public audiences in mind. As mentioned earlier, in order for students to become familiar with the power of language and rhetoric and see that their words also have impact in discussions of environment and place, assignments and writing tasks must be provided which encourage students to write for audiences other than their teachers. We agree with Paul Heilker: "Writing teachers need to relocate the *where* of composition instruction outside the academic classroom because the classroom does not and cannot offer students real rhetorical situations in which to understand writing as social action" (71). Assignments should encourage students to examine and participate in public debates in the places where they live and explore and write about those issues. These assignments should be context driven and provide students with room in which they can direct their writing toward issues of both local and global concern.

One of the most compelling aspects of ecocomposition pedagogy is its links to Paulo Freire's dialogic methodology. That is, ecocomposition, like Freire's pedagogy, asks students to participate in conversations with both their environments and other members of their community or biosphere. When Freire offers his problem posing approach to gaining hold of local discourses and local literacies, he is, in essence, asking members of a community to question the very roles of their environments and to engage those environments in dialogue. Dialogue places members of a community, such as writing students, *within* writing environments rather than asking them to merely write *about* those environments. In other words, as we see it, ecocomposition—à la Freire—asks students to write in their environments, to be critical of those environments, and to consider what effect their own writing and literacies have on those very environments. It seems that what Freire was getting at, and what ecocomposition should be getting at, is that if we ask students solely to write *about* environment we situate both student writers and ourselves outside of environment, creating an oppositional position, one which might lend to dominant, oppressive views of those environments. But, encouraging writers to be critically aware of their environments and to participate in them situates writers as participants who affect and are affected by those environments, rather than merely observers with a detached curiosity about those places. For example, an assignment about local clean up efforts might read:

> It seems that cleaning up the places in which we live has become an agenda of almost all American communities. Most likely the community in which you currently live or where you grew up (should they be different places) is home to several organizations which orchestrate volunteer cleanups of local environments. Such efforts are extremely beneficial to local communities, but in order for them to be effective, volunteers must be reached and encouraged to participate. Volunteer work can be provided in many ways.
>
> One thing that you might consider is volunteering your skills as a writer to produce publicity material or fliers announcing upcoming events and encouraging new volunteers to participate.
>
> *Assignment:* choose a local organization whose agenda meets with your own environmental goals. Design a flier which details the agenda and affiliations of the organization and invites volunteers to participate in an upcoming event.
>
> Remember that you will be writing for a very diverse audience, but an audience which will have some similarities in how they think about local environments (some one apathetic to cleaning up local rivers, for instance, is not likely to read a flier about a clean-up project), so you can make certain assumptions about your audience. You will need to be convincing, since motivating some concerned individuals to active participation may be difficult.

A second assignment might also read:

> Quite frequently, readers come to learn about organizations which they did not previously know existed through reading what others have written about those organizations. When writers portray agendas of organizations, frequently those writings are the only access readers have to an organization. Hence, a writer's words have real effect on readers' understandings of political agendas of environmental organizations. Similarly, writers frequently provide the only access many readers have to understanding local environmental legislations. For instance, newspaper columnists often present editorialized articles which express whether or not a piece of legislation is beneficial to a local community. These articles often sway readers' ways of thinking about that very issue.
>
> There are countless forums through which to inform readers about local issues and organizations. One of the more effective ways is electronically: the World Wide Web and the Internet.

Assignment: Select an organization or issue which is particularly interesting to you. Develop a Web page which explains, in detail, the organization's goals, the way to join the organization, the details of the legislation, and so on. Provide links to other similar pages so that your readers may access even more information about this subject.

Frequently the argument is made that students, particularly first-year composition students, do not possess the authority to publish their writing. Such a view not only oversimplifies the notion of publication, limiting the definition of publication to widely read journals, magazines, books, and newspapers, but also denies students' rights to participate in public discourse. That is to say, academics too often think of public writing as published writing, and we tend to assign academic standards of value to how and what we evaluate and teach students regarding public writing. However, as the assignments provided above exemplify, students can find public outlets for their writing; teachers must begin to explore broader definitions of publication so that they might direct their students to a variety of public places for their writing. Encouraging students to produce written work which reaches readers beyond their classrooms encourages students to value their own writing as part of larger, public conversations. Such encouragement helps students to better understand the power of rhetoric and writing while at the same time giving students real reasons to learn the "subject matter" of the issues they've chosen to address and support. As publisher/author Nick Lyons has written about public writing and activism in relation to rivers that need protection, "words can help. Public words are often the only way to galvanize such support. And the words need not only be the strident, if well-meaning shouts of outrage, but the words of affection and understanding" (62).

While we stand by our belief that the writing instructor's first duty is to teach writing, we also believe that students learn to write when they are writing to learn. Effective discourse is *engaged* discourse; students learn best when they are invested in the subject about which they write. In other words, it is the teacher's responsibility to ensure that students are placed in situations where their writing can reach real audiences for real purposes. Learning more about a particular subject—whether that subject is the natural environment, the political struggles involved in environmental debates, or any other subject of importance—is inseparable from learning to write effectively and well. Discursive growth can happen only when a writer is engaged with a subject that is worthy of engagement. To our thinking, learning about our global, national, and local environments is perhaps the most important subject of study today. We agree with Jagtenberg and McKie when they suggest that the "nonhuman physical environment . . . is so central to sustainable life that it undermines the very idea of space and the biophysical world as a context for human activity" (xii). Quite simply, writing, like other human activities, requires an environment that will support life. Perhaps significant of the need to immediately begin to incorporate ecological literacy approaches to pedagogy—both composition and otherwise—is this headline from CNN that ran while we were writing this book: "A baby boy—symbolically designated the world's 6 billionth person—was born to a first-time mother in a Sarajevo hospital Tuesday, the head of a pediatric clinic said. Dr. Idris Bukvic said Fatima Nevic gave birth to the 8-pound boy two minutes after midnight. The U.N. Population Fund had estimated the world's population would reach 6 billion on Tuesday, and Secretary-General Kofi Annan said he would declare the first child born in the Bosnian capital after midnight Baby Six Billion." (CNN Headline News, Online, October 12, 1999). The report continued to say that "A September report by the UNPFA estimates that by 2050, some 8.9 billion people will be living on the earth." As frightening as it was when McDonald's advertising boasted "over six billion served" many years ago, the fact that the United Nations is advertising "over six billion born" is even more frightening, considering our world's depleting natural resources.

Ecological Discourse Approach

The second type of ecocomposition pedagogy—which we have called the "ecological discourse approach"—asks students to not only consider environmental issues as subjects about which to write and think, but also asks students and teachers to consider the very ecologies of writing. Once again, when we speak of environments, we mean the entire spectrum of places, locations, and positions from which discourse arises. Ideally, such an approach to teaching writing should be combined with the ecological literacy approach in order to ask students to look more holistically at the relationships between written discourse, their own writing, and environments and environmental issues that affect their lives. In other words, while ecological literacy-based approaches may be developed devoid of the ecological discursive approach and vice versa, we offer that a more encompassing ecocomposition pedagogy involves both.

The ecological discourse pedagogy is situated within the notion that words, language, and writing are themselves parts of ecosystems and that when writers write they affect and are affected by environment. That is, like the general definition which says that ecology is the study of the relationship between organisms and their environments, this pedagogy examines the activity of writing and its relationship to surrounding discursive acts and locations. This pedagogy asks that we, as teachers, conceptualize writing not as an individual activity, separated by the author from the world, but an activity of and in the world. It asks that we step beyond examining the process of individual writers to examining larger environmental forces on those writers. It asks that in addition to the ideological, cultural contexts in which we have situated writers in recent times, that we look to physical environments, textual relationships, and the locations from which language and discourse arises. It asks us to see writing as an activity of relationships.

Writing happens through connections, interactions, and relationships. The very act of composing a sentence consists of a number of ecological situations. First, we recognize the ecology of interaction between language users uttering (or writing) the sentence, those attempting to understand it, and the locations in which these interactions occur. Without such interaction, meaning cannot take place. Meaning-making requires others. As Thomas Kent argues, "Without the other, we can have no thoughts, no language, no cognizance of meaning, and no awareness that we possess something called mental states" ("Production," 304). In addition, the subjects, thoughts, or ideas that are being negotiated arise from a long history of previous interactions between innumerable language users. While we aren't suggesting a "Progress Narrative," we do argue that all knowledge—and each sentence—is dependent upon an ecological history of thoughts and ideas. Our current knowledge—à la Kuhn—responds to and reacts upon previous acts of knowledge-making. Importantly, these interactions occur in particular locations (and also build upon interactions in previous locations) and are shaped and transformed by them. We also recognize the relationship between words in sentences, sentences in paragraphs, and so on. Each word or sentence or paragraph has meaning only through its relationships with other words, sentences, and paragraphs. Language, communication, knowledge, and writing are all ecological pursuits.

In his essay "Writing Takes Place," Sid argues that "Writing is an ecological pursuit. In order to be successful, it must situate itself in context; it must grow from location (contextual, historical, ideological). . . . Writing does not begin in the self; rather, writers begin writing by situating themselves, by putting themselves in a place, by locating within a space" (18). Quite literally, writing begins with place, with *topoi*. From that place, writers then turn to a host of relationships with other texts, other writers, other places in order to write.

A particularly good example of the ecological discourse type of pedagogy are assignments that employ hypertextual writing—commonly known today as "webbed writing." Writing in hypertextual formats seems to be inherently ecological. Whereas conventional methods of writing reinforce the idea of the singular author writing in a heirarchical, linear style, writing on the World Wide Web facilitates associations, connections, and junctures between words, ideas, texts, and authors. These connections are manifested through links, cyclical references, and paths of information that are not often found in traditional texts. A hypertext has no canonical order. In a sense, webbed writing projects often work like ecosystems. Ideas, words, and texts connect with one another, work relationally by referencing one another, and assume no intrinsically heirarchical order of importance. The use of hypermedia, such as digital images and sounds, adds another dimension to webbed writing, allowing for more diversity and variety—the attributes of a healthy ecosystem. That is, hypertext writing insists that writers pay attention to the environment in which they write, not just the writing.

Scholars including Ernst Curtis, J. M. Gellrich, and more recently, Jay David Bolter have traced similarities and analogies between the concepts of knowledge, ideas, and the world of nature, suggesting that our conceptions of how knowledge is organized often correspond to our conceptions of the natural world. For example, during the Middle Ages, the world was viewed as a great encyclopedia of information to be ordered, classified, and controlled. Consequently, books themselves were seen as the instruments through which humans might most effectively and accurately undertake this regulation. It is no accident that printed books and encyclopedias of the time were exceedingly heirarchical and ordered. In fact, Gellrich suggests that the underlying ambition of most thinkers, theologians, and encyclopedists of the Middle Ages was "to gather all strands of learning together into an enormous Text, an encyclopedia or summa, that would mirror the historical and transcendental orders just as the Book of God's Word (the Bible) was a speculum of the Book of his Work (nature)" (18).

The advent of electronic books and texts has the potential to break down some of these hierarchical conceptions of nature and text. Webbed writing can indicate a hierarchy of topics, but its basic structure does not inherently categorize and codify information in the same linear structure as does a traditional printed book. Writing hypertextually reflects a mindset that more accurately corresponds to our current conceptions of ecology. Jay David Bolter writes eloquently on this subject:

> The electronic book reflects a different natural world, in which relation-ships are multiple and evolving: there is no great chain of being in an electronic world-book. For that very reason, an electronic book is a better analogy for contemporary views of nature, since nature is often not regarded as a hierarchy, but rather as a network of interdependent species and systems. The biological sciences dispensed with the great chain of being over a century ago—long before the advent of the electronic computer. More recently, but also long before the computer, physics rejected simple hierarchical views of matter and energy. In fact the metaphor of the book of nature has long been moribund. But with the coming of the computer, we have a writing technology that suits a contemporary scientific conception of the world, and the metaphor of the world as a hypertextual book can now be explored. We can expect contemporary scientists and scholars to come more and more to the conclusion that the book of nature is a hypertext, whose language is the computational mathematics of directed graphs. This is an intriguing prospect. For if scientists are studying the interdependencies of nature, while humanists are reading hypertexts, then our vision of nature can be reunited with our technology of writing in a way that we have not seen since the Middle Ages. (105–6)

Ecocomposition seeks to contribute to this reunion between scientific conceptions of ecology and ecological conceptions of discourse and text—in both theory and practice.

Students easily grasp the ecological nature of the webbed writing environment, and hypertextual writing assignments, when coupled with environmentalist discussions and texts, often seem to reinforce one another. The study of natural ecologies and the foregrounding of "ecological" types of discourse work together to underscore to central tenet of ecological thinking: that environments (whether they are natural, discursive, or otherwise) consist of interactions and relationships. For example, as a part of a WAC initiative at University of Tampa, Christian taught an ecocomposition course that was linked with a Global Issues course focusing on environmental issues. The students met with Christian and their Global Issues professor at different times, but the courses shared several texts, discussed related and overlapping issues, participated in joint activities with both professors, and completed several shared assignments. In many senses, the very coupling of these two courses was an ecological, sustainable, interdisciplinary endeavor of the sort that Derek Owens argues for. For their final writing assignment of the semester, the students worked together in groups of three to develop group web projects. These projects focused on environmental and ecological issues on campus.[4] After composing a homepage for the project as a class, the students selected the following issues for group study: recycling on campus, energy use, flora and fauna on campus, water treatment and wastewater management, and waste disposal and hazardous waste management. The students quickly discovered that the university, situated just outside of downtown Tampa, was inextricably linked with the city, the county, and the surrounding environments. The project led students to the fact that what they do on campus affects the larger ecosystems in the area, including the Hillsborough River and Tampa Bay, and in turn, the Gulf of Mexico, and ultimately other marine environments. In addition, the students quickly perceived connections between each others' web projects, and by including links and references to one another's work, the entire project became much more ecological.

By the project's completion, Christian's students had created an ecosystem of texts and ideas, including links to local, national, and global environmental groups and organizations, references to each other's work, and graphics and digital images of the campus and its surroundings. As part of one group's project, they initiated a cleanup of the campus park, and gave special thanks to each of the other groups for participating. After numerous peer-evaluation workshops, the students presented their work to an interdisciplinary faculty panel. Nearly every group referenced the others, and most came to see the entire web project as something of which they each owned equal shares. One student, during her group's presentation, stated that this "web project allowed us to connect with each other much like plants and animals do in an ecosystem." Webbed writing assignments of this sort are much more ecological, in that they allow students to see bits of information, pages, links, ideas, and so forth relationally. Students often recognize that knowledge depends upon its relationships with other knowledge, that facts, texts, and even selves depend upon shared resources, and that productivity can often be something group generated and group maintained. In that respect, we feel that hypertextual writing reinforces ecological, environmental thinking more than traditional linear forms of writing which seem to reinforce individualistic, self-centered thinking.

The ecological discourse approach to ecocomposition is not, however, bound to hypertextual assignments. The ecological discourse approach asks that we encourage students to engage discourse as an ecological activity. Discourse does not begin in the self, as some expressivist theories and pedagogies have erroneously suggested; rather, writing begins externally in location. Writers write by situating themselves, by locating themselves in a particular space/context. The *what, why,* and *how* of each act of writing is inextricably linked to the *where*. As Edward M. White points out: "The very word topic comes from the Greek word for 'Place,' suggesting that the thinking process is a kind of geographic quest, a hunt for a place where ideas lurk" (8). In writing from a place, a *topoi,* writers write that place, define that place. Writers engage in a circular reinscription of place and environment that in turn writes who they are as writers. Teaching students to be cognizant not only of the words they write on the page, but also of the environments from which those words grow and in turn influence, helps students to understand the larger relational importance of writing and also helps them to discover that writing does not occur non-contextually, that writing affects living.

An oft heard mantra of the writing teacher is "context and convention." Context is the situated place where writing happens. Not just the physical environment where a writer writes, but the environment of writing, the ideological environment, the cultural environment, the social environment, the economic environment, the historical environment. Context is the interrelationship between words that give meaning to text. Context is environment, not just the environment where writing takes place, but the environment where words are situated in relationship to other words, to other knowledges, to other texts, to other traditions in order to construct a system of words that have meaning. Much like organisms that may stand alone only for short periods of time, words require relationships with other words, with their environments in order to function. Words and ideas are dependent upon other words and ideas for survival. Conventions of writing are nothing more than the ordering of words into relationships which provide for sufficient survival in particular contexts, particular environments. Context, as all writing teachers know, dictates convention. The activity of writing is the activity of creating an ecosystem of words. Sentences, paragraphs, texts are habitats for words, for knowledge. When we talk about continuity and coherence in a paragraph, we're talking about symbiotic relationships between inhabitants of a paragraph. Coherence is symbiosis. The first thing ecocomposition teaches us is to know your terrain; identifying context as environment allows writing to find its place in an environment more effectively. Teaching writing ecologically situates writing; it does not, like some writing pedagogies, simply teach writing as a subject, something to be examined and practiced outside of context, outside of environment. Ecocomposition identifies that all writing happens not only contextually, but environmentally. Education theorists C. A. Bowers and David J. Flinders lend toward this type of pedagogy when they write that the classroom "must be understood as an ecology of language processes and cultural patterns" (*Responsive*, 2). We encourage writing instructors at all levels to employ ecological models of discourse into their own pedagogies. Understanding the acts of writing and writing courses as ecosystems allows both students and teachers to see discourse as an activity of connectedness.

CHAPTER 6

Ecocomposition:
Perspectives, Perceptions, and Possibilities

꒞ꑦ꒥

We discover that we are constantly in dialectical relationships with our environments.
—Victor Villanueva, Jr., "Considerations for American Freireistas"

To know the spirit of place is to realize that you are a part of a part and that the whole is made of parts, each of which is whole. You start with the part you are whole in. —Gary Snyder, *The Practice of the Wild*

To be in relation to everything around us, above us, below us, earth, sky, bones, blood, and flesh, is to begin to see the world whole, even holy.
—Terry Tempest Williams, "The Erotic Landscape"

The theory of place does not simply propose a binary separation between the "place" named and described in language, and some "real" place inaccessible to it, but rather indicates that in some sense place is language, something in constant flux, a discourse in process.
—Bill Ashcroft, Gareth Griffiths, and Helen Tiffin,
The Postcolonial Studies Reader

I thought again about throwing language all over a scene, wondered if the emotional mystery of one's response to place doesn't lie in the inchoate play of possible words, of felt meanings and poetries, of the sublime, the romantic, the picturesque, Zen; even, perhaps, something new. And perhaps that twinge of disappointment one always feels at the words chosen—and thus also at the glorious scene—comes from the dream that in that instant of indecision and all-decision before your mind clarified its response to beauty, you just might have held within you language finally saturated with all the earth's meaning.
—Daniel Duane, *Caught Inside: A Surfer's Year on the California Coast*

We hope we have made it clear that ecocomposition is an endeavor that addresses the most focal issues in rhetoric and composition studies today while also negotiating and extending the very boundaries of the discipline. Ecocomposition examines discourse as a system of relationships that is embedded in, and also significantly shapes and structures, nearly all natural, constructed, and social environments. Discourse, like life itself, is dependent upon relationships, affiliations, and associations. We agree with Fritjof Capra, who writes that "an outstanding property of life is the tendency to form multileveled structures of systems within systems" (28). We would go further to suggest that discourse is one of the most significant of these systems, since it affects nearly every aspect of life as we know it and how we know it. By drawing on other disciplines' conceptions of systems and relationships, ecocomposition advances vitally new approaches and ideas to the study of discourse.

Ecocomposition suggests a number of compelling theories that envision discourse ecologically. It argues that discourse creates systems, both literally and figuratively, believing that an essential property of all systems is communicative relationships among its members. Whether we're talking about the Hawaiian coral reef system, plants and animals in the Florida Everglades, or the readers and authors of *College English,* all of the members of a system operate relationally, communicating and interacting with other members. In other words, discourse is that which defines and delineates a system in that members of a community often speak a discrete "language" that separates them from other communities. Alligators and academics alike speak in their own languages, and how and what they say defines who and where they are. Paradoxically, discourse also offers the means through which various communities might successfully interact with other communities. Discourse can draw disparate groups closer to communication, resolution, and mutually beneficial interaction. In fact, through discourse we may find that no real separation actually exists between communities, that all groups and individuals share common resources, desires, and ways of being. However, we do want to note that discourse is also the vehicle through which relations of oppression, hegemony, and dominance are manifest. We do not wish to suggest that ecocomposition's approach to discourse and ecology offers a harmonious vision of relational existence. Rather, ecocomposition looks to understand how relationships to environments and other organisms might contend with oppressive hegemonies, and we do not wish to suggest that ecocomposition offers utopian promises, only critical gazes. As human beings, each of us is a member of many different groups, ecosystems, and discourse communities, and the language we use at any given moment can never emerge from one exclusive location or environment. In other words, we come from multiple discursive ecosystems, and what we think and say is the product of all of these locations, despite their seeming dissimilarities. James E. Porter offers a useful discussion of this relationship between discourse and ecology in his book *Audience and Rhetoric: An Archaeological Composition of the Discourse Community:*

> Discourse communities may operate a little like ecosystems. An ecosystem is a convenient ecological space defined by certain characteristics that set it off from abutting systems. But shift your perspective slightly and the borders of the original ecosystem break down, because ecosystems inevitably interact with systems abutting them. Discourse communities cannot be isolated any more than the writer can be isolated as an object of study from his social field. In other words, we need to remember that discourse communities overlap—and are flexible and locally constituted. They may cross academic and institutional boundaries, and they may exist only momentarily. (86)

As a relatively new, yet thriving and diverse ecological system itself, ecocomposition contributes to the study of discourse in potentially momentous ways, and this contribution owes much to ecocomposition's attention to studies of discourse and environment across many disciplines. We have identified a number of disciplines that have been significant in ecocomposition's short history, including ecology, environmental studies, sociobiology, philosophy, ecofeminism, ecocriticism, cultural studies, linguistics, communication, literary studies, and of course rhetoric, and composition. We hope that ecocomposition will draw from an even greater number of disciplines and perspectives in the future.

Perhaps the most important lesson we can learn from these interdisciplinary studies is that important and unforseen ideas often arise through the coupling of seemingly unrelated disciplines such as composition and ecology. The junctures between disciplines can yield more than a simple recognition of mutual and overlapping of interests; they are often the birthplaces of new ways of thinking, writing, and teaching. Interdisciplinary endeavors like ecocomposition have the potential to generate what the philosopher C. D. Broad called "*emergent properties*"—properties that emerge at a certain level of complexity but do not exist in their separate states (Capra, 28). Such thinking, often applied to work in ecology and biology, suggests that the nature of the whole is something different from the sum of its parts, and unprecedented characteristics, traits, and insights often arise through interaction and combination. The taste of sugar, as we noted earlier, is not present in the separate elements of carbon, hydrogen, and oxygen; the essence of sugar emerges only when these separate elements are combined. The same holds true for ecocomposition; the theories and pedagogies that have been produced in this subdiscipline are distinctive and unique from work that has been produced in any other singular discipline or approach. Ecocomposition in its present form probably could not have emerged without the intermingling of a variety of perspectives and disciplines. It is our hope that ecocomposition will produce work of even greater "sweetness" in the future.

Also significant is the close relationship between environmentalism, both as an ethical position and as an area of intellectual inquiry, and ecocomposition. As we have noted, ecocomposition owes much to various perspectives on the natural world. Ecocomposition's birth is linked to both nature writing and environmental discourse, and as such, we do not expect that it could be—or should be—separated from these perspectives. Not only does discourse shape our human interactions with each other, it also affects the earth itself. As the (current) dominant species on this planet, the language we use for and about it will have a significant impact on its future. The worlds of biosphere and semiosphere come together in ecocomposition, and from this confluence we learn much about both of these spheres. Furthermore, rhetorical studies of environmental discourse—produced by individuals and groups from a wide range of institutional and cultural perspectives—are among the most topical and important subjects that rhetoricians might address today. No other discursive subject in the public sphere is as wide-ranging and ubiquitous as environmentalism. Environmental rhetoric is, as Herndl and Brown suggest, "so varied in fact that it connects almost every part of our social and intellectual life, crossing the boundaries between various academic disciplines and social institutions" (4).

The biosphere and the semiosphere are mutually dependent upon each other, and they shape and affect one another in important ways. From ecology and environmental studies, we learn much about systems of organization. From environmentalist perspectives, we come away with a deeply ethical sense of our relationship with the world. And from composition and discourse studies, we learn how all of these perspectives are mapped, negotiated, constructed, and defined through language—in essence, how they are written. However, although ecocomposition began as an attempt to bring studies of the natural world into the composition classroom, we hope it will evolve to account for more holistic investigations of the relationship between place, space, environment, and discourse. As we've suggested, discourse affects and is affected by nearly all locations. The class-room, the public sphere, the Internet—all of these are imbedded in and with discourse. As such, they fall within the realm of ecocomposition. What is most important is that ecocomposition continues to foreground the *production* of discourse and its relationship to environment, for such work is not only important, it is essential.

While the previous chapters have focused, to some degree, on the his-tory and context of the relationships between discourse and ecology, we'd like to conclude by looking ahead at a few of the many important areas of inquiry in ecocomposition's future. To use a geographic metaphor—as we so often do in this book—ecocomposition is a journey just begun, and the path ahead is as yet unknown. While we cannot say where our collec-tive travels will lead us, we can begin to imagine something of what we might see along the way. As an ecological endeavor itself, we foresee changes and developments in ecocomposition coming about from within and without. Certainly, new theories, pedagogies, ideas, insights, and dis-cussions will germinate from the work of the many committed ecocom-positionists working in this discipline. As writing teachers undertake new ecologically tuned classroom activities, they will surely develop new peda-gogical insights. In addition, as ecocomposition scholars and theorists converse and interact with one another through conferences, journals, and email lists (like the newly formed ASLE-CCCC listserv and the ASLE-CCCC Special Interest Group), we expect that new and meaning-ful ideas will emerge to cultivate and advance ecocomposition theory. We also expect that those same teachers and scholars, as well as academics working in related disciplines, will find other productive intersections between the study of discourse and the study of environments. In short, we feel that ecocomposition will develop into a diverse, multifaceted field of inquiry, and we anticipate a flowering of activity (sorry, we can't avoid the eco-metaphors at this point) that could potentially give rise to one of the most significant and constructive approaches to the study of discourse in the years to come.

In this chapter, we take up a number of issues and ideas that we feel will become more important as ecocomposition evolves and matures in the fu-ture. Perhaps one of the most important of these are theories and ideas that loosely revolve around *the personal*. We argue here that the self, subjectivity, and expressions of personal experience and positionality are vitally impor-tant to ecocomposition. The way we refashion both our personal and cul-tural beliefs concerning ecosystems—discursive, natural, or otherwise—will be of direct relevance to the worlds we build in the future. Self-expression has the potential to add breadth and diversity to ecocomposition, and all that we identify as expressions of the self and the subjective will contribute to determining our ability to deal with difference, otherness, and multi-plicity. Social constructivist approaches often rely too heavily upon collec-tivist, cumulative perspectives and risk overlooking the individual's posi-tion and experience and how those are shaped through interaction with various environmental factors. We address ecocomposition's appreciation of the individual and individual experience, and we also recognize the im-portance of the individual in the activity of triangulation. By viewing the communicative situation as a relationship between self, other, and envi-ronment, ecocomposition identifies discourse as an ecological endeavor.

We also address a topic closely related to the personal—the emotional aspects of rhetoric and discourse. Ecocomposition recognizes that emotion is an important component of humans' reactions to environments. While this is exemplified through the often impassioned uses of discourse in na-ture writing texts, the emotional need not be limited to purely "natural" environments. Ecocomposition argues that the study of discourse must pay more attention to emotional responses to environments. Toward this goal, we turn to classical rhetoricians, who acknowledged and fore-grounded the ecocentric, emotional aspects of discourse *(pathos)* and iden-tify a strong link between individual agency and environment *(nomos-physis)*. An emotional approach to the relationships between discourse and environment seeks to locate human values and ethics in a harmonious re-lation to our environments. A more complete (and ecological) understand-ing of how language and environment interact depends upon a greater rec-ognition of the personal, the emotional, and the subjective uses of discourse.

As should be obvious by now, ecocomposition draws heavily upon the-ories, methodologies, and subjects in the sciences—particularly the natu-ral sciences including ecology, environmental science, biology, and the like. However, ecocomposition hopes to move beyond the rationalistic, mechanistic world view of Descartes, Newton, Bacon, and their scientific and methodological progeny in support of a more integrated, holistic, ec-ological world view. We suggest that greater attention to the personal, the emotional, and the ethical moves us closer to an integrative world view. To be clear, it is important to distinguish between an attention to the *per-sonal* and an overemphasis of the *self-assertive*. Recognizing that personal experiences and ethical and emotional aspects of discourse are vital and important in the support of diversity is a great deal different than suggest-ing that self-assertion and domination are the principal characteristics of life. In other words, we feel that the discourse we produce and extol can be attuned to personal, subjective perspectives without being overly self-interested, domineering, or grounded in the neoromantic. Many Western social and economic structures are fundamentally rooted in an anti-ecological, exploitive paradigm (what Riane Eisler calls the "dominator system" of social organization) which advocates the control and consumption of resources. We feel—as do many postmodern thinkers—that this paradigm is limiting, inadequate, and often flatly immoral. The incorporation of personal, emotional perspectives as well as cooperative, integrative approaches moves ecocomposition—and all other areas of academic and intellectual inquiry—closer to a more holistic, ecological, balanced world view. As Capra writes in *The Web of Life:* "If we now look at our Western industrial culture, we see that we have overemphasized the self-assertive and neglected the integrative tendencies. What is good, or healthy, is a dynamic balance; what is bad, or unhealthy, is imbalance—overemphasis of one tendency and neglect of the other" (9). We conclude this chapter by calling for more distinct yet wide-ranging examinations of the relationships between discourse and environment, including investigations of computer-generated environments, specific public locations of discourse, and imagined and theorized sites of language use.

Emotion

Like the culture/nature binary that has permeated Western thinking, so too have reason and emotion been set in opposition. Ecocomposition, in its resistence to the culture/nature binary—in fact, in support of its argument that culture and nature are inseparably enmeshed—also contends that reason and emotion cannot be separated when examining issues of environment and place. That is, ecocomposition contends that first-hand experience is crucial to one's understanding of interaction with a place and that personal reaction and response to that experience is critical and valid to the interpretation and expression of experience. In fact, we turn to neurobiology in support of the belief that emotion is an inseparable part of the thought process. Recent neurological research identifies that the Cartesian belief in a "split" between the mind and the body is a fallacy, and that the release of chemicals that accompany instinctual, emotional responses often and significantly alter the thought process. Such thinking calls into question the separation between emotion and logic, seeing both as integral, related ways of thinking. In fact, several neurologists have recently argued that emotion and logic, at least in terms of neurochemical activity, are nearly indistinguishable. Antonio Damasio, reporting on this neurological research in *Descartes' Error: Emotion, Reason, and the Human Brain,* writes that "the action of biological drives, body states, and emotions may . . . be an indispensable foundation for reason" (200). From this perspective, it could be argued that the absence of emotion in Western thought is perhaps even more pernicious than the often-feared excess of emotion.

As ecocompositionists, we do not see emotion as a deep reflection of some inherent "self," but rather as individual, learned responses to particular stimuli. That is, emotions are part of the constructed self, a self partially constructed by the environment in which that self grows, a self discursively formed. We do not wish to forward a conception of emotion grounded in romanticism, but rather an idea that emotive responses are valid, important responses that help us to understand our reactions to particular experiences. However, we do believe that each person's emotional response is unique, in that it arises from a complex history of that person's previous experiences. Since no two individuals, no matter how similar, could ever share the exact same experiences, each individual expresses his or her own unique perspective of each particular situation. What is most important about this uniqueness is that each situation or experience will be viewed differently by each person; complete objectivity is impossible. Each situation is interpreted and expressed through the individual's own emotionally biased perspective. As Val Plumwood notes in her 1991 *Hypatia* article "Nature, Self, and Gender: Feminism, Environmental Philosophy, and the Critique of Rationalism," Western thinking puts forth a notion of reason and emotion as

> sharply separated and opposed, and of "desire," caring, and love as merely "personal" and "particular" as opposed to the universality and impartiality of understanding and of "feminine" emotions as essentially unreliable, untrustworthy, and morally irrelevant, an inferior domain to be dominated by a superior, disinterested (and of course masculine) reason. This sort of rationalist account of the place of emotions has come in for a great deal of well-deserved criticism recently, both from its implicit gender bias and its philosophical inadequacy, especially its dualism and its construct of public reason as sharply differentiated from and controlling private emotions. (5)

While emotion has existed in some contexts in Western thinking, it has been mostly relegated to the creative arts. As such, it stands in sharp opposition to the rationalistic thinking that long dominated the sciences. In a sense, scientific rationality has been defined as that which "gets things done in the world," while emotion and the arts are seen as something superfluous and, at best, life enhancing. Ecocomposition must move beyond the traditional Western dismissal of emotion and look to personal reactions and responses as critical when engaging all environments. Ecocomposition recasts the personal and emotional as a dialectic, highlighting the importance of "particular" perspectives in order to avoid totalizing narratives and metanarratives. Ecocomposition also acknowledges that there is no personal without the social, that all knowledge, including the emotive, derives from interaction, collaboration, and triangulation. In other words, emotion is the unique response of the individual in a given situation, and that response is the product of a plethora of previous interactions, collaborations, and dialogues with others. Most simply put, then, emotion is a constructed response, not a reflection of an inherent self.

Talking about emotion in relation to concerns of place and environment can be a difficult task. Frequently, the emotive response is disregarded as unbalanced or irrational and aligned with the feminine; hence, it is devalued. Similarly, individuals and groups aligned with environmental concerns are often cast as too emotional (and therefore incapable of making sound, rational decisions or judgments) by debating opponents. As a result, they are easily disregarded. Ecocomposition must work to amend this belief while also recognizing the predominance and authority of scientific, rationalistic discourse in many, perhaps most, discourse communities in the West. In other words, ecocomposition must be attuned to the relationships between emotion and science, between "knowledge" and emotive response, and it must not allow either to dominate, but instead encourage the two to tussle, react, enmesh, and engage.

We also suggest that an emotional response to a particular moment is an important part of the experience. Emotion is not an unrational response; it is a facet of how our constructed selves react based on previous learned responses to stimuli. According to Edward O. Wilson, emotion is the

> modification of neural activity that animates and focuses mental activity. It is created by physiological activity that selects certain streams of information over others, shifting the body and mind to higher or lower degrees of activity, agitating the circuits that create scenarios, and selecting ones that end in certain ways. The winning scenarios are those that match goals preprogrammed by instinct and the satisfaction of prior experience. (123)

Like the triangulation model of communication we explained in chapter 2, Wilson's understanding of emotion is one of learned response, one in which prior experiences trigger certain passing responses. As Wilson explains, "current experience and memory continually perturb the states of mind and body" (123). In other words, like Thomas Kent's prior and passing theories, Wilson acknowledges that our prior experiences and our never-ending passing experiences (which become prior experiences) determine how we react emotionally. That is to say, just as triangulation argues that knowing the world occurs through matching one's interpretations with another about the world, so too is emotional response a matter of matching responses with others and the world. For the most part, emotion is merely another reactive form of communication, one based on prior experiences and as valid and important as any "thought-out" response. As Wilson claims, "without the stimulus and guidance of emotion, rational thought slows and disintegrates. The rational mind does not float above the irrational; it cannot free itself to engage in pure reason" (123).

Wilson also turns to Antonio Damasio's neurological research into the foundations of human emotions. For example, Wilson notes Damasio's identification that there are two broad categories of emotion. The first, what Damasio calls "primary emotions" are instinctive neural responses, what those who study animal behavior often call "releasers." According to Wilson, in humans, these responses include "sexual excitement, loud noises, the sudden appearance of large shapes, the writhing movements of snakes or serpentine objects, and the particular configuration of pain associated with heart attacks and broken bones" (125). Primary emotions have not changed much throughout the evolution of humans and likely existed in prehuman vertebrates as well.

Secondary emotions are those that Wilson explains as developing from individual experience. What makes the difference between primary and secondary emotions interesting is that they are exhibited physiologically in the same ways. Nature, as Wilson quotes Damasio, "with its tinkerish knack for economy, did not select independent mechanisms for expressing primary and secondary emotions. It simply allowed secondary emotions to be expressed by the same channel already prepared to convey primary emotion" (125). We find it particularly interesting that "natural" primary emotions and learned secondary emotions manifest themselves in the same manner. For ecocomposition, such a neurological insight informs the bound relationship between nature and learned behavior. That is, because secondary emotions are based partially on environments in which they are learned, and primary emotions transcend environment, but are both regulated and transmitted from the same location, it seems to suggest a greater link between socially constructed learning and "natural" human knowledge. In other words, the exhibition of secondary emotions is merely another communicative display of learned response. The fact that secondary emotions are conveyed physiologically through the same manner as primary emotions suggests to us the inability to cast human beings as separate from the biological, chemical world and asks us to recognize humans as intricately made up of biological and chemical systems as well as being parts of larger systems.

For ecocompositionists, then, it stands that emotional response offers a particular set of information about experiences. Hence, when issues of environment (natural and otherwise) are examined discursively, rationality must not be the only vehicle through which analysis is performed. As Steven B. Katz and Carolyn R. Miller have noted in "The Low-Level Radioactive Waste Site Controversy in North Carolina: Toward a Rhetorical Model of Risk Communication," emotional reaction is "sometimes a more appropriate and reasonable response than logic" (131). For Katz and Miller, in fact, emotion becomes an important aspect of environmental rhetoric particularly when there is no certainty about scientific data used in an argument. For Katz and Miller, as for other ecocompositionists and environmental rhetoricians,

> emotions, like values, are an integral part of decision making insofar as they inform human perception, understanding, and communication; emotions motivate and underline all human knowledge and behavior, and thus any attempt to ignore, suppress, or exclude them in the decision-making process can result in misunderstanding between parties to the decision. (131)

In his essay "Epistemic Responsibility and the Inuit of Canada's Eastern Arctic: An Ecofeminist Appraisal," Douglas F. Buege also identifies the importance of emotion in what he calls "responsible knowing." According to Buege, responsible knowing "focuses upon individual people as knowers who have a responsibility to obtain and use knowledge in activities in which they participate and are accountable for that knowledge" (103). Included in Buege's five points of responsible knowing is an acknowledgment of the importance of emotion. He argues that "responsible knowing takes emotions as central to cognitive practice. Emotions are an essential component of oppressed people's knowledge of the world" (104).

Like Buege and others who have emphasized the importance of emotive responses in the decision making process, we too are eager to see ecocomposition readily acknowledge the importance of emotion, particularly when addressing issues of environment and ecology. However, what is most interesting to us as ecocompositionists is the mixing of "natural" emotive responses with learned, constructed emotive responses. That is, emotion is often set in binary opposition to rationality specifically because rationality is seen to exhibit control, whereas emotion represents that which is uncontrolled. Much like the ecocolonial conquests of "wild" environments, emotion has often been cast as the domain of the feminine, and hence been historically another body over which to exert control. A move to include emotive response as integrally crucial to understanding issues of environment is a move to resist the control wielded over natural bodies, landscapes, and environments. And in doing so, it is an attempt to better understand why and how norms and nature have been constructed and accepted. That is, the emotive gives way to exploring what we have accepted as "natural" versus what we have constructed as correct. In other words, in our look to the future possibilities for ecocomposition, we contend that we must first reconsider the ways in which we have constructed phenomena that occur "naturally." For instance, as teachers of writing, we are often intrigued by the ways in which composition studies has (for the most part) accepted concepts of "the writing process" or of particular genres as accepted ways of doing things. The writing process, in fact, is often cast as the "natural" way to engage the activity of written production. Only recently have post-process theorists begun to call into question the "natural" order of the writing process. Similarly, genres and rhetorical modes are taught as conditions which exist as "natural" forms of writing—as evidenced by the many composition textbooks that support and propagate this view. Of course, few compositionists would argue that any genre is Natural, rather than constructed. Yet, the question of what conditions of writing exist "naturally" versus those that have become norms lead ecocomposition to ask about the very relationship between the natural and the accepted norm.

Ecocomposition and the *Nomos-Physis* Antithesis

As we've mentioned previously, nearly all of Western thinking begins as a meditation on humans' relationship to the Earth. In their search for the foundations and essence of human life, early Greek thinkers looked for an order in living systems around them which exemplified the meaning, direction, and purpose of all life. Perhaps the most noteworthy and important aspect of the early Greeks is the broad inclusiveness of their methodology. In their attempts to make sense of their world, they incorporated scientific, philosophical, and spiritual thinking in their understandings of life. As David Macauley writes, "In their search for a hidden *arche,* and underlying *logos,* or a guiding *telos,*" early Greek thinkers "looked to the growth, movement, and relations of plants and animals. Science at this point was virtually indistinguishable from philosophy" (1). In other words, Western thinking was ecological from the outset.

As an integral component of early Greek thought, it is no surprise that the study of discourse and rhetoric contained similarly ecological perspectives. For instance, the concepts of *nomos* and *physis*—forwarded by the Sophists in the later part of the fifth century B.C.E.—have much in common with our current ecological conceptions of discourse. Simply put, *nomos* are accepted ways of doing things—norms, if you will. According to Susan Jarratt, *nomos* may have initially meant "pasture," but evolved to take on the meaning "a range of words," particularly as it was used metaphorically in Homeric epic (41). Later, the word's meaning moved from pasture to habitation, "signifying," according to Jarratt, "'habitual practice, usage or custom'" (41). Jarratt continues to explain that "Common to both forms is the importance of human agency: in the first case, in the marking out and distributing of land; and in the second, in explicitly human ratification of norms as binding" (41).[1] That is, the move to understand *nomos* as the place one inhabits suggests a comfort, a familiarity, an acceptance of conditions. That is, the Sophists began to identify *nomos* as meaning something that is "believed in, practiced, or held to be right" (Guthrie, 55). Those things practiced and "held to be right" became guidelines on which decisions and policies (both social and political) were based. In other words, conditions that were "right," that were habitual practices, began to inform decision making processes. *Nomos,* according to W. K. C. Guthrie "presupposes an acting subject—believer, practitioner, or apportioner—a mind from which the *nomos* emanates" (55). Hence, there can be no one *nomos,* but instead "different people had different *nomoi*" (55). Different people accepted different conditions and practices as the "right" habitual practices, the right way to do things, the right way to make decisions. The right way to do things, the *nomos,* was frequently cast as antithetical to the ways things "Naturally" occur, to *physis.* That is, conditions of the norm were viewed not as natural, but as contrary to nature.

Physis is generally understood to mean "nature." *Physis* are those conditions and characteristics that occur not by choice, not because they are "right" or because they have been decided upon, but simply because they occur "naturally." According to G.B. Kerferd, *physis* was also the term that "the Ionian scientists came to use for the whole of reality. . . . But it also came to be used to refer to the constitution or set of characteristics of a particular thing, or class of such things, especially a living creature or a person, as in the expression 'the nature of man'" (111). *Physis* seems to suggest a difference between "natural" characteristics enjoined upon a thing and those characteristics adopted and appropriated by the thing. The term often lends itself toward the understanding of things as being the way they are "because they have grown or become that way" (Kerferd, 111). Guthrie also attributes *physis* as simply meaning "reality" (55).

Hence, *nomos* and *physis* are set in opposition to one another, dividing the *nomos,* what has been conceived and thought to be right from the *physis,* that which is natural, preordained, or congenital. Or as Jarratt explains it, "rhetoric can be closely linked with *nomos* as a process of articulating codes, consciously designed by groups of people, opposed both to the monarchical tradition of handing down decrees and to the supposedly nonhuman force of divinely controlled 'natural law'" (42). For our purposes the relationship between *nomos* and *physis* are critical because of what using the *nomos-physis* antithesis poses as inquiry. As Guthrie writes, when the Sophists began to question the roles of *nomos* and *physis:*

> Discussions of religion turned on whether gods existed by *physis*—in reality—or only by *nomos;* of political organization, or whether states arose by divine ordinance, by natural necessity or by *nomos;* of cosmopolitanism, on whether divisions within the human race are natural or only a matter of *nomos;* of equality, on whether the rule of one man over another (slavery) or one nation over another (empire) is natural and inevitable, or only by *nomos;* and so on. (57–58)

For ecocompositionists, this sophistic question is critical as we see the beginnings of asking as to whether nature and environment are conditions of *physis* or *nomos.* Of course, we would contend that *nomos* and *physis* are inextricably bound and that what is natural is also a condition that has come to be accepted. However, what is more important to us as ecocompositionists is what the *nomos-physis* antithesis stands to teach us about discourse.

Discourse, discursive functions, constructions, communities all operate as *nomos.* That is, discourse is not a "Natural" condition, but is a series of norms, or *nomos,* prescribed as the accepted way of doing things. If we identify discursive conventions as a series of *nomos,* then we also have to identify that they operate as antithetical to *physis,* to that which is natural. There, then, can be no "natural" characteristics of discourse. Yet, discourse, like *physis,* can acquire characteristics much the same way that which is *physis* can exhibit assumed traits. Thus, discourse as a system transcends the *nomos-physis* antithesis and stands as a system that is rhetorically both *nomos* and *physis.* In other words, the *nomos-physis* binary is a discursive means by which to taxonomize and codify natural from constructed, and discourse becomes an entity that resists that very codification. That is, ecocomposition's turn to *nomos* and *physis* is not simply a move to identify that the Sophists understood a difference between rhetorically constructed *nomos* and "Naturally" existing *physis,* but is a move to bring the two terms together as enmeshed and necessary conditions of each other. Discourse— that illusive system which can be both "natural discourse" and rhetorical— is the very vehicle which binds the human agency of *nomos* and the "nature" of "human nature" which was of such interest to the Sophists. Simply put, ecocomposition looks to identify connections between natural environments and rhetorical environments. Ecocomposition has the possibility to teach much more about discourse, about writing, by identifying that writing conventions, such as genres, do not exist as *physis,* but as *nomoi,* though we often teach them as "natural" or correct ways to produce writing.[2] In other words, when we begin to examine writing ecologically, we begin to see the characteristics that have grown to be accepted as both *physis* and *nomos.* The obvious example of this is the ways in which genres have evolved to reflect *nomos* of particular writing cultures while also maintaining characteristics of *physis* (though no writing can ever be truly located in the *physis*—only perceived as such).

Aristotelian Rhetoric

While this investigation of *nomos* and *physis* can be quite useful to ecocompositionists in their studies of discourse, early Greek thought offers other examples of ecological thinking about discourse. Perhaps the most widely read, studied, and practiced investigation of discourse in any discipline or time period is Aristotle's *On Rhetoric.* Aside from his role as one of the first important botanists, biologists, and naturalists, Aristotle was also one of the first important philosopher/rhetoricians. In fact, it could be argued that the act and art of discourse in the Western world has been more influenced by Aristotle's fourth century B.C.E. text than by any other. What is most interesting for ecocompositionists is that what he endorses in *On Rhetoric* is in

essence an ecological conception of discourse. For example, Aristotle's trea-
tise envisions persuasive discourse as transcending particular discourse
communities; in a sense, Aristotle was perhaps the first thinker to examine
the relationships between various disciplines of what are now the arts and
sciences. He argues that rhetoric "does not belong to a single defined genus
or subject but is like dialectic" (35). Aristotle goes on to define rhetoric as

> an ability, in each particular case, to see the available means of persuasion.
> This is the function of no other art, for each of the others is instructive and
> persuasive about its own subject: for example, medicine about health and
> disease and geometry about the properties of magnitudes and arithmetic
> about numbers and similarly in the case of the other arts and sciences. But
> rhetoric seems to be able to observe the persuasive about "the given," so to
> speak. That, too, is why we say it does not include technical knowledge of
> any particular, defined genus of subjects. (37)

For the first time in Western thinking, discrete disciplines were emerging
as separate areas of study. Although Aristotle saw differences and divisions
between these disciplines, he took a more holistic view of the role of rhetoric in shaping and structuring them; consequently, he looked at their
boundaries as flexible, relational, and connected—all through discourse.
For example, George A. Kennedy suggests that

> Although Aristotle largely limits the province of rhetoric to public address,
> he takes a broader view of what that entails than do most modern writers on
> communication. This often surprises and interests readers today. He ad-
> dresses issues of philosophy, government, history, ethics, and literature; and
> in book 2 he includes a comprehensive account of human psychology. (ix)

In other words, Aristotle envisioned all disciplines relationally, imagining
the study of discourse as that which both encompasses and distinguishes
them. In a sense, Aristotle's view of disciplinarity is not so different from
that endorsed by contemporary thinkers like Wilson and Capra, although
rather than seeing the hard sciences as the "glue" which adjoins the disciplines, Aristotle looks instead to rhetoric and discourse. Certainly this is a
view that is attuned to ecocomposition—or more precisely, a view to
which ecocomposition should *become* more attuned.

Moreover, Aristotle's conception of the communicative situation is ecological. Aristotle, more than rhetoricians before him, and more than many
rhetoricians following him, understood communication as an ecology of
speaker (or writer), subject, and listener. He suggests in book 1 that "a

speech situation consists of three things: a speaker and a subject on which he speaks and someone addressed" (47). While his discussions and classifications of human psychology in book 2 of *On Rhetoric* seem formulaic and stereotypical to us today, they nevertheless show Aristotle's view that communication consists of a relationship or connection between at least two individuals speaking on a topic in a particular location. His attention to the audience's mental state—be it one of anger, calmness, or friendliness—shows his concern for more than just the rhetor. Aristotle seems to suggest that "speakers need to understand how the minds of their listeners work, and in the process we come to understand something of who we are and why we do what we do" (Kennedy ix). In addition, Aristotle argues that seeing the available means of persuasion *(pisteis)* involves a complex understanding of *ethos, pathos,* and *logos*—the familiar triad of speaker's character, audience's emotion, and the logic or proof of the subject or argument itself. Effective discourse requires an interplay of these three means of persuasion, requiring the rhetor to determine the amount and style of each according to the situation. In effect, this view suggests that the discursive situation requires diversity; only through the interactions between ethos, pathos, and logos can effective—or healthy—discourse emerge.

Seen as a whole, Aristotle's view of communication is both diverse and contextual, suggesting that each discursive situation is an ecosystem of ideas, arguments, characters, and topics. The capacity to discover the "right thing to say" in a given situation—*phronesis*—suggests that rhetors need to be able to read and understand the environments in which their discourse will exist, and the degree to which their utterances will survive depends on this ability. One wouldn't last very long in the Florida Everglades without some knowledge of and preparation for that particular environment; similarly, discourse has very little chance of survival if the rhetor does not first learn about and prepare for its entrance into a particular discourse community. Since Aristotle is often accredited with being the first of many professions—the first botanist, for example—and since his rhetorical work certainly has affected ecocomposition as well as composition studies in general, perhaps it is not too bold of us to say that Aristotle was the first ecocompositionist.

Rhetorical Appeals: Past and Present

We've mentioned here the importance of ethos, pathos, and logos to classical rhetoricians such as Aristotle. We've also mentioned the importance of recognizing the ethical, moral, and emotional aspects of discourse of all sorts. For ecocomposition, these topics converge most significantly in some of the recent research and scholarship being done in environmental rhetoric. In their investigations of the rhetoric used by environmentalist organizations, naturalists, legislators, and corporations, environmental rhetoricians have been among the first to recognize the impact and frequency of emotional and ethical appeals in public discourse about the environment. Environmental rhetoricians all agree that environmental discourse is a historically developed cultural form that is both constructed and maintained by rhetorical activity, and many of these scholars offer models of analyzing environmental discourse that can be quite easily traced to Aristotle's original discussions of rhetoric.

For example, Killingsworth and Palmer's *Ecospeak* offers a "Continuum of Perspectives on Nature" which attempts to map some of the most prevalent and distinct ethical and epistemological perspectives on nature. The authors offer their continuum in an effort to move beyond the simple dualism of environmentalist versus developmentalist, and they quite rightly suggest that such an effort might begin to "break the hold of ecospeak by identifying various discourses on the environment before they are galvanized by dichotomous political rhetoric" (10). While Killingsworth and Palmer make it clear that their mapping of environmental discourse is somewhat oversimplified and schematic, we find their model most useful in its attempt to "plot the major perspectives along a continuum whose poles designate three human attitudes toward the natural world" (11).

The authors suggest that these three attitudes—nature as object, nature as resource, and nature as spirit—correspond with the primary perspectives or subject positions involved in environmental debates in the public sphere. At the extreme perspective is traditional or mainstream science, with its concern for objectivism and its "fabled detachment from all natural objects," which sees nature as an object to be rationalized, classified, and defined (12). Obviously, such a perspective relies heavily on a Logocentric

Nature as Object		Nature as Resource		Nature as Spirit	
traditional or mainstream science	government	business & industry	agriculture	social ecology (Humanistic Environmentalism)	deep ecology (wilderness ethic, nature mysticism)

CONTINUUM OF PERSPECTIVES ON THE ENVIRONMENT (Reprinted, by permission from Killingsworth and Palmer, *Ecospeak*, 11.)

means of persuasion that promotes the logic or proof of the subject or argument itself. In other words, mainstream science, in both its epistemology and its public rhetoric, asserts a quantifiable, analytic, logical perspective on nature—an appeal to *Logos*. In the center of Killingsworth and Palmer's continuum is the viewpoint of nature as a resource, which is primarily adopted and advocated by business, industry, and agriculture. This perspective relies heavily on the belief that nature is a bounty of resources for human use and enjoyment; however, the rhetoric of this perspective (particularly in recent years) often asserts the importance of ethical use, conservation, and management. In this viewpoint—the rhetoric of the marketplace—the concept of "ethical action is retained but applied exclusively to human behavior, with the rest of the world reduced to meaningless motion" (13). In other words, businesses, agricultural organizations, and industries espouse an anthropocentric viewpoint of balance between proper use in the present while maintaining a commitment to preservation for the future—an appeal to *ethos*. At the opposite pole are environmentalists and deep ecologists, who embrace the viewpoint of nature as spirit. This perspective asserts a "mythic involvement with nature, an identity in which the spirit of creation wraps the human and the nonhuman in an indissolvable unity" (12). This third viewpoint places humans on a par with nature and involves "an active resistance to the other perspectives that violate that identity of human beings and nature" (13). In their rhetorical attempts to make the unity between humans and the earth more clear and profound, environmentalist groups often depict environmental degradation in human emotional terms, showing the earth as a suffering being. In fact, deep ecologists work to close the gap between science and environmentalism by turning toward "radical" scientific theories such as James Lovelock's Gaia hypothesis (the concept of the Earth as literally a self-regulating living organism). Clearly, such a perspective works primarily on an emotional level—an appeal to *pathos*.

So, in a very basic sense, Killingsworth and Palmer's analysis of the primary subject positions in environmental discourse corresponds to the primary rhetorical appeals as first analyzed by Aristotle in *On Rhetoric*. The various "players" in environmental debates all work largely from one rhetorical perspective or another. As should be clear, the emotional appeal is far from absent in environmental debates. In fact, the appeal to *pathos* has been adopted by those occupying one of the most vocal and pronounced subject positions involved in public discussions of the natural world—environmental activists and deep ecologists. Killingsworth and Palmer's analysis supports the claim that emotional appeals are an important aspect of dis-

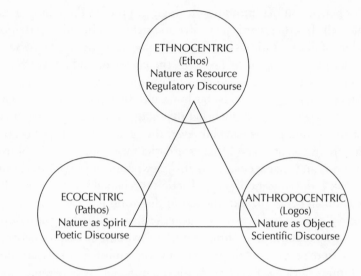

A RHETORICAL MODEL FOR ENVIRONMENTAL DISCOURSE (Reprinted, by permission from Herndl and Brown, *Green Culture*)

course about the environment; *pathos* is shown as a particularly important aspect of the rhetoric of ecotopian fiction and poetry, which the authors address in chapter 6 of their book. However, as Killingsworth and Palmer suggest, effective rhetors in environmental debates often employ a complex mixture of all of the various attitudes toward nature listed in their continuum. Moreover, these same rhetors often "seek to achieve a measure of control over an audience or an opponent in debate by categorizing the opponent into a single role assigned on the basis of a dominant attitude" (12). In other words, those involved in public debate about nature often portray themselves as having a breadth of attitudinal experience while claiming that their opposition works from a sole (and therefore limited) perspective. By limiting their opposition's perspective to one isolated viewpoint—and the primary persuasive appeal that corresponds to it—these rhetors are better able to present their own perspective as more balanced and equitable.

The introduction to Herndl and Brown's *Green Culture* also attempts to organize analysis of environmental rhetoric. Their model, based in part upon Killingsworth and Palmer's "Continuum of Perspectives on Nature," is designed to "identify the dominant tendencies or orientation of a piece of environmental discourse" (10).

Herndl and Brown present a triangular rhetorical model for environmental discourse, and they place "Regulatory Discourse" at the top of the model. The authors suggest that regulatory discourse—the discourse of powerful institutions that make decisions and set environmental policy—usually regards nature as a resource to be managed for the greater social welfare. The rhetorical power of this discourse "emerges from the rhetorical notion of *ethos,* the culturally constructed authority of the speaker or writer who represents these institutions" (11). At the second point of their rhetorical triangle of environmental discourse, Herndl and Brown place the specialized discourse of the environmental sciences. Within this discourse, nature is often regarded as "an object of knowledge constructed through careful scientific methodology" (11). According to the authors, the rhetorical power of this discourse emerges from "the notion of *logos,* the appeal to objective fact or reason" (12). At the third point of their triangle, they place the poetic discourse of nature writers. This discourse asserts that nature holds a spiritual or transcendental beauty, and writers like Henry David Thoreau, Aldo Leopold, Rachel Carson, and John Muir represent the view that humanity is a part of nature. The power of this discourse "comes largely from aesthetic or spiritual responses to the notion of pathos, or appeals to the emotions" (12).

Herndl and Brown emphasize, as do Killingsworth and Palmer, that these discursive characteristics are not pure and only represent the dominant tendencies of the key players in environmental debates. As the essays in their collection attest, successful environmental discourse "often combines the styles, forms, and rhetorical appeals of more than one of these discourses" (12). Herndl and Brown quite usefully identify the three primary rhetorical appeals—naming them specifically in their study—that are used in debates about nature, a fact which further emphasizes the degree to which the study of environmental rhetoric can be traced to studies of rhetoric in ancient Greece. Obviously, the appeal to pathos is far from absent in current discussions about nature, and ecocomposition would do well to recognize its importance to its own related studies. While we've mentioned the connections between ancient Greek rhetoric and current environmental debates only specifically in reference to these two major works, studying the wealth of other rhetorical inquiries into environmental discourse could yield even further correlations between the two.

Ecocomposition's Future

We hope that by now we have made clear some larger characteristics of ecocomposition and what ecocomposition can become. Without question, ecocomposition must concern itself not only with "Natural" environments, but with all environments. Yet, at the same time, ecocomposition must acknowledge and work toward alleviating the "environmental crisis" that threatens the very life systems of this planet. In order to achieve these goals, ecocomposition, both as a theory for research and as a pedagogy, must be active. That is, it must engage local environments. Ecocompositionists themselves must be active. As Sid suggests in "Writing Takes Place," "ecocomposition is not a term for definition, but an inquiry for action." Ecocomposition must consider the role of publics, the classrooms in which it is taught, and the emotions of individuals as they encounter any and all environments. And, at the same time, ecocomposition must expand its own range of tolerance to explore other areas of relationships between discourse and environment.

For instance, we see computer environments, cyberspace, and hypertextual discursive locations as ripe for ecocomposition to explore. Certainly, the relationships between the imagined spaces of cyberspace, the discourse used to create those locations, and the discourse used within those places must be further theorized and considered. For example, ecocomposition must be concerned with what happens when MOO space is created and MOO users "build" rooms and objects to occupy those rooms through discursive mapping. Similarly, increased development of "Networked Writing Environments" at universities around the country lends to examining the relationships between writers, writing technologies, and electronic writing environments. But ecocomposition must also move beyond looking merely at computer environments specifically designed for writers and students of writing. Ecocomposition must also look at "cyborg writing," at the work Donna Haraway and others are doing concerning the relationships between technology and life. We must explore the relationships between human organisms who rely on technology for survival and the spaces they occupy in a biotechnological world. As we become more and more dependent on technology in all human activities—perhaps most profoundly in communication—we must consider the ramifications of these new technological spaces on our lives.

Likewise, ecocomposition must extend its examination of the role of public discourse in electronic and other spaces. We must further consider what public space *is*. Much of the current work in rhetoric and composition involving "the public" deals almost exclusively with the actors, writers, and speakers who are involved in public discourse. Drawing largely from work in sociology, history, cultural studies, and other anthropocentric fields of study, rhetoric and composition's current understandings of public discourse consistently undertheorize and overlook the environments and places in which such discourse occurs and the extent to which public conversations are shaped, altered, and constructed by environment itself. Ecocomposition must work to further theorize the role of environments in public discourse. For many ecocompositionists, this will mean a greater attention to the ways that public discourse shapes and constructs *natural* environments. Again, while we encourage work of this sort, we also hope to forward a more holistic notion of the relationship between discourse, be it public or private, and environments, be they natural or constructed.

Ecocomposition must also consider what role our own discourse will have in future public conversations about environmental issues. Clearly, we must reconsider what it means to be an activist intellectual, and we must use our roles within and outside of the academy to speak with and for underrepresented groups and places. We might combine our personal *pathos* with our institutional *ethos* and *logos* to urge others to work for the preservation of natural places. As eco-activist intellectuals, we might adopt a range of eco-strategies, working with diverse groups and audiences. We might also use our roles as educators to advance public understanding on issues relating to the natural world. While we maintain that ecocomposition must explore a wider range of environments and their relationships with discourse, we also recognize that all environments, all discourse, and all human and nonhuman activity depends upon a larger life-supporting environment that we call Earth. This environment requires our immediate and diligent efforts, and we must draw upon all of our institutional and intellectual resources to help maintain and preserve it. According to Stephen Jay Gould:

> We cannot win this battle to save species and environments without forging an emotional bond between ourselves and nature as well—for we will not fight to save what we do not love. . . . So let them all continue—the films, the books, the television programs, the zoos, the little half-acre of ecological preserve in any community, the primary school lessons, the museum demonstrations, even (though you will never find me there) the 6:00 A.M. bird walks." (*Eight Little Piggies,* 40)

Eco-activism works with and beside other important social movements in support of diversity, difference, and equality. Environmental concerns have been given particular attention within the women's movement, indigenous peoples' struggles, and other types of progressive social activity, and composition's devotion to issues of race, class, and gender might be extended to include natural environments as well. Ecocompositionists are particularly well-equipped to address issues of discourse and environment, both within the academy and in larger public conversations, and it is imperative that we use our talents for the appreciation and preservation of natural places. One way in which ecocompositionists might address these issues in the classroom is to move from using multicultural-based readers or other issues readers and take up issues of ecological and environmental concerns in the classroom in the same ways other critical issues have been taken up. Certainly, the textbook market now provides a range of good composition-oriented textbooks that can assist in this endeavor.

Final Thoughts

Simply put, what ecocomposition does and must continue to do is explore the co-constitutive relationships between discourses and environments. In doing so, ecocomposition must make uncomfortable natural discourse and see discourse and nature alike: neither "Natural," both constructed, and both reliant upon and affecting the other. In order to do so, ecocomposition should take as its primary agenda the study of the relationships between all environments and the production of written discourse. We cannot emphasize enough that production, not interpretation, must be the cornerstone of ecocomposition.

Having said that, let us end here not with concluding remarks, but rather with prompts and proddings to continue ecocomposition investigations. In chapter 1, we offered that the left side of our title was designed to make uncomfortable the relationships between nature and discourse in order that we not accept discourse as natural, but rather we begin to explore the inextricably entwined relationships between discourse and nature. We'd like to turn now to the right side of that same title and consider what it might mean to move "toward ecocomposition." We have argued that composition studies is already deeply vested in ecological inquiries—inquiries that question the relationships between writers, texts, knowledge, discourse, culture, race, gender, and a host of other categorical labels. Similarly, we have pointed out how composition might benefit from turning to methodologies employed by ecological sciences, and we have inquired as to what role all environments might play in writing and, in turn, what role writing might play in constructing environments. These are the very sorts of inquiries we hope to see ecocomposition move toward.

It is our hope that as composition studies continues to expand its focus on writing processes, on the social conventions of writing, on cultural issues of writing, and on questions of disciplinary boundaries, that composition now move toward ecocomposition, that composition begin to explore the importance of the role of place, environment, space, site, location in ways that allow composition studies to both account for the importance of those places and to understand their affect on written discourse. Ecocomposition has the potential to guide composition studies to exciting new sites of inquiry—presuming, of course, that we are willing to accept ecocomposition's link first to the natural world and then move beyond the limited scope of seeing ecocomposition as being about nature only. Ecocomposition is about all places and their relationships to discourse. Composition cannot ignore the crucial role that all environments play in the production of discourse, nor can critical studies of environments ignore the role of discourse in constructing them. Quite simply, discourse and environment cannot be separated.

Notes

Chapter 1: Ecocomposition

1. Throughout this text, we use the terms *environment, place, nature,* and *natural* in various ways. As we elaborate, *nature*—and, in turn, *place* and *environment*—are socially constructed. That is, "Nature" is mapped discursively. However, in order to distinguish between this social construct and those parts of the world which are not "(hu)man-made," we use the adjectival description *natural* in order to delineate moments when the assumption should be invoked that systems other than human systems are responsible for such creations. In other words, the tag word *natural* should indicate a sort of erasure under which we present items as existing beyond human construct. Much as Nietzsche did with the term *nihilism,* we do so only as a means by which to talk about "natural" formations and retain an underpinning understanding that such moments may only be accessed through socially constructed consensus. Only when the term *nature* appears in quotes and capitalized ("Nature") should it be read under the same erasure. The only exception to this is when the word appears within a quoted reference; in such cases we have left the presentation of the word as the original author presented it.

Chapter 2: The Evolution of Ecocomposition

1. For more on Marilyn Cooper see also Erika Lindemann, "Three Views of English 101." (*College English:* 57.3, 1995).
2. We recognize the danger in referring to something as "Darwinian" as Darwin argued for many elitist, racist, sexist notions of who in the human race ought to be allowed to survive. In *The Descent of Man,* and *Selection in Relation to Sex,* for instance, Darwin argues that individuals who are poverty stricken should be prohibited from marrying and reproducing. In essence, Darwin argues that only the elite be allowed to reproduce in order to ensure a higher order of human survival. Such proclamations in the name of evolution and in the name of science are abhorrent and not to what we refer when we identify Cooper's model as Darwinian. Rather we use the reference to refer to the popular views of "natural selection," "evolution," and "survival of the fittest," to which Darwin is accredited.
3. See Dobrin, "Paralogic Hermeneutic Theories, Power, and the Possibility for Liberating Pedagogies," in *Post-Process Theory: New Directions for Composition Research,* 145–83.

Chapter 3: Ecology and Composition

1. For an excellent history of the development of writing in ancient cultures, see Diamond, "Blueprints and Borrowed Letters," in *Guns, Germs, and Steel: The Fates of Human Societies*, chapter 12.
2. For a detailed look at the Greek understanding of ecological harmony, see Egerton, "Ancient Sources for Animal Demography," 175–89.
3. Keep in mind that the purveyors of writing and of knowledge about "Nature" were primarily men. Hence, it is important to recognize the eco-feminist critique of how such knowledge was managed and how it was used to colonize and oppress.
4. Interestingly, the redfish is known by a number of different names in different parts of the country. In Florida, it is called the Redfish, while other geographic regions know it as the Red Drum, the Red Bass, the Channel Bass, or the Drum. Nature is constructed and named differently in different geographic locations and by different communities; living creatures, perhaps more than any other discursive topic, are prone to various names accorded by various peoples.
5. Exile to Kansas is not fun for creatures who are best-suited for aquatic climes; turtles don't like it, neither did Sid whose range of tolerance is similar to that of the Kemp's Ridley.

Chapter 4: Ecocomposition and Activist Intellectualism

1. For a more detailed explication of public writing, see Christian's *Moving Beyond Academic Discourse: Composition Studies and the Public Sphere* (Carbondale: Southern Illinois University Press, 2002).
2. See Worsham, "Writing against Writing: The Predicament of *Ecriture Feminine* in Composition," in *Contending with Words*, 82–104, and Dobrin, *Constructing Knowledges* for critiques of this tradition.
3. Granted, such tourism does begin the process of awareness. Many would argue that any such introduction to public writing will initiate a willingness to participate in public writing and will introduce students to the conventions of public writing. We would agree that this is true in principle; however, the letter to the editor assignment often operates outside of many public writing conventions as it allows most anyone to say most anything with little if any public response. Similarly, many would argue that our example of the artificial rainforest experience, like the artificial public writing experience, is indeed an actual experience and that the simulation is as real as that which is represented. Of course, we would agree with this, but we have some deeper resistance to brief moments of tourism as substitute for hands-on experience.
4. As of this writing the CCCC Service Learning web site had not yet been developed, but NCTE recommends the following resources to learn more about service learning: in formation about the NCTE service learning e-mail conversation list can be found at http://www.ncte.org/ch/#subscribe; Tom Deans' web site at http://www. personal. ksu.edu/~tdeans/csl.html; and the Campus Compact site at http://www.compact.org/.
5. Often professors were the cause of the problem too.
6. University presses such as SUNY, Oxford, Harvard, Utah, Virginia, Georgia, Nevada, Kentucky, Arizona, Idaho, and Illinois, as well as Island Press, Seal Press, to name but a few.
7. The magnitude of this debate and the far-reaching implications are only glossed here. The S.O.S. debate has drastically changed how environmental policy is discussed and

enacted in Florida. In fact, four years following the passing of the S.O.S. legislation, gubernatorial candidates' positions on S.O.S. were still being scrutinized.

Chapter 5: Ecocomposition Pedagogy

1. Though we use the word *system* here to describe the concept of writing process, we recognize the importance of post-process writing theories which contend that there can be no codifiable system of discursive production or interpretation. However, as most post-process theorists contend, such an understanding of discourse production also makes teaching writing difficult, if not impossible. Hence, composition pedagogies—to essentialize—are often derived from a process approach to writing, one in which writing processes are taught as controllable, measurable systems. In order to more easily approach ecocomposition pedagogy, we will account for composition pedagogies as part of the process paradigm, though ecocomposition pedagogies will push at the edges of those systems. For more about post-process theories, see Kent, *Post-Process Theory,* p.

2. The Florida state system centralizes all courses. That is, in order for a course to be approved in a state school, it must be approved at the state level and given a state number. For instance, all state colleges and universities identify first-year composition—as ENC 1101. The Gordon Rule requires that in each course labeled a writing course that a particular number of written words must be assigned and assessed in order for that course to meet the Gordon Rule requirement. For example, ENC 1101 is listed as a 6,000 word Gordon Rule course; in such courses students must be required to write and instructors must assess a minimum of 6,000 words of writing. The Gordon Rule was imposed by the state legislature.

3. At the time we are writing this Derek Owen's manuscript is being revised for NCTE. We are grateful for his willingness to share the unpublished manuscript with us. Similarly, our collection is due for release from SUNY Press, and so our references here refer to the unpublished manuscript versions of the contributions.

4. For a more extended definition of ecocomposition and webbed writing assignments turn to Bradley Monsma's "Writing Home: Composition, Campus Ecology, and Webbed Environments" in our collection *Ecocomposition: Theoretical and Pedagogical Approaches,* pp. 281–290.

Chapter 6: Ecocomposition: Perspectives, Perceptions, and Possibilities

1. Interestingly, Susan Jarratt does not pair *nomos* in opposition to *physis* as did many of the Sophists and as do Guthrie, in *The Sophists,* and Kerferd, in *The Sophistic Movements.*

2. Bawarshi "Beyond Dichotomy," 69–82. See his work in genre theory for an extended discussion of genre as *nomoi.*

References

Abbey, Edward. *Desert Solitaire: A Season in the Wilderness*. New York: Touchstone, 1968.
———. *The Monkeywrench Gang*. New York: Avon Books, 1975.
Abram, David. "Merleau-Ponty and the Voice of the Earth." In *Minding Nature: The Philosophes of Ecology*. Edited by David Macauley, 82–101. New York: Guilford Publications, 1996.
Adler-Kassner, Linda, Robert Crooks, and Ann Watters, eds. *Writing the Community: Concepts and Models for Service-Learning in Composition*. Urbana, IL: NCTE and AAHE, 1997.
Anderson, Chris, and Lex Runciman, eds. *A Forest of Voices: Conversations in Ecology*. Mountainview, CA: Mayfield Publishing Company, 2000.
Ashcroft, Bill, Gareth Griffiths, and Helen Tiffin. *The Post-Colonial Studies Reader*. New York: Routledge and Kegan Paul, 1995.
Aristotle. *Historia Animalium*. Trans A. L. Peck, Vol 1–3. Cambridge, MA: Harvard University Press, 1979.
Aristotle. *On Rhetoric*. Trans. George A. Kennedy. New York: Oxford University Press, 1991.
Arnold, David. *The Problem of Nature: Environment, Culture and European Expansion*. Cambridge, MA: Blackwell, 1996.
Bawarshi, Anis. "Beyond Dichotomy: Toward a Theory of Divergence in Composition Studies." *JAC: A Journal of Composition Theory* 17 (1997): 69–82.
Bennett, Tony. "Putting Policy into Cultural Studies." In *Cultural Studies*. Lawrence Grossberg, Cary Nelson, and Paula Treichler, eds. New York: Routledge and Kegan Paul, 1992.
Berland, Judy. "Angels Dancing: Cultural Technologies and the Production of Space." In *Cultural Studies*. Lawrence Grossberg, Cary Nelson, and Paula Treichler, eds. New York: Routledge and Kegan Paul, 1992.
Berlin, James A. *Writing Instruction in Nineteenth-Century American Colleges*. Carbondale: Southern Illinois University Press, 1984.
———. *Rhetoric and Reality: Writing Instruction in American Colleges, 1900–1985*. Carbondale: Southern Illinois University Press, 1987.
———. *Rhetorics, Poetics, and Cultures: Refiguring College English Studies*. Urbana: NCTE, 1996.
Berlin, James A., and Michael J. Vivion. *Cultural Studies in the English Classroom*. Portsmouth, NH: Boynton/Cook, 1992.
Biehl, Janet. *Rethinking Ecofeminist Politics*. Boston: South End Press, 1991.
Birkerts, Sven. *The Gutenberg Elegies: The Fate of Reading in an Electronic Age*. New York: Fawcett Columbine Books, 1995.

Bizzell, Patricia. "Patricia Bizzell's Statement, " http://www.hu.mtu.edu/cccc/98/social/ bizzell.htm

Bolter, Jay David. *Writing Space: The Computer, Hypertext, and the History of Writing.* Hillsdale, NJ: Erlbaum, 1991.

Bookchin, Murray. *Ecology of Freedom.* Palo Alto, CA: Cheshire Books, 1982.

Bowers C. A. *Critical Essays on Education, Modernity, and the Recovery of the Ecological Imperative.* New York: Teachers College Press, 1993.

———. *Education, Cutlural Myths, and the Ecological Crisis: Toward Deep Changes.* Albany: State University of New York Press, 1993.

———. *Educating for an Ecologically Sustainable Culture: Rethinking Moral Education, Creativity, Intellegence, and Other Modern Orthodoxies.* Albany: State University of New York Press, 1995.

Bowers, C. A., and David J. Flinders. *Responsive Teaching: An Ecological Approach to Classroom Patterns of Language, Culture, and Thought.* New York: Teachers College Press, 1990.

Bowler, Peter J. *The Norton History of The Environmental Sciences.* New York: W. W. Norton, 1992.

Brewer, Richard. *Principles of Ecology.* Philadelphia: Saunders, 1979.

Bruffee, Kenneth. "Collaborative Learning and 'The Conversation of Mankind.'" College English 46 (1984): 635–52.

———. "Social Construction, Language, and the Authority of Knowledge: A Bibliographical Essay." College English 48 (1986): 773–89.

Bryant, Paul T. "Nature Writing: Connecting Experience with Tradition." In *Teaching Environmental Literature: Materials, Methods & Resources.* By Frederick O. Waage. 93–101. New York: MLA, 1985.

Buege, Douglas F. "Epistemic Responsibility and the Inuit of Canada's Eastern Arctic: An Ecofeminist Appraisal." In *Ecofeminism: Women, Culture, Nature.* Edited by Karen J. Warren, 99–111. Bloomington: Indiana University Press, 1997.

Buell, Lawrence. *The Environmental Imagination: Thoreau, Nature Writing, and the Formation of American Culture.* Cambridge, MA: Harvard University Press, 1995.

Bullard, Robert D. *Dumping in Dixie: Race, Class, and Environmental Quality.* Boulder, CO: Westview Press, 1994.

Burke, Kenneth. *Permanence and Change: An Anatomy of Purpose.* 2d ed. Indianapolis: Bobbs-Merrill, 1965.

Capra, Fritjof. *The Web of Life: A New Scientific Understanding of Living Systems.* New York: Anchor Books, 1996.

Carson, Rachel. "*New York Herald-Tribune* Book and Author Luncheon Speech." In *Lost Woods: The Discovered Writing of Rachel Carson.* Edited by Linda Lear, 76–82. Boston: Beacon Press, 1998.

Chodorow, Nancy. *The Reproduction of Mothering: Psychoanalysis and the Sociology of Gender.* Berkeley: University of California Press, 1978.

Coe, Richard M. "Defining Rhetoric—and Us: A Meditation on Burke's Definitions." *Journal of Advanced Composition* 10 (1990): 39–52.

Cokinos, Christopher. "What Is Ecocriticism?" Http://wsrv.clas.virginia.edu/~djp2n/conf/ WLA/cokinos.html. 9/23/97 9:44am.

Collett, Jonathan, and Stephen Karakashian. *Greening the College Curriculum: A Guide to Environmental Teaching in the Liberal Arts.* Washington, D.C.: Island Press, 1996.

Cooper, Marilyn, and Michael Holzman. *Writing as Social Action*. Portsmouth, NH: Boynton/Cook, 1989.

Crockett, Harry. "What is Ecocriticism?" Http://wsrv.clas.virginia.edu/~djp2n/conf/WLA/crockett.html. 9/23/97 9:44 am.

Cushman, Ellen. "The Rhetorician as an Agent of Social Change." CCC 47 (1996): 7–28.

Damasio, Antonio R. *Descartes' Error: Emotion, Reason, and the Human Brain*. New York: Avon Books, 1995.

Darier, Eric, ed. *Discourses of the Environment*. Malden, MA: Blackwell, 1999.

Diamond, Jared. *Guns, Germs, and Steel: The Fates of Human Societies*. New York: W. W. Norton, 1999.

Dixon, Terrell. "Inculcating Wildness: Ecocomposition, Nature Writing, and the Regreening of the American Suburb." In *The Nature of Cities: Ecocriticism and Urban Environments*. Edited by Micahel Bennett and David W. Teague, 77–90. Tucson: University of Arizona Press, 1999.

Dobrin, Sidney I. "Writing Takes Place." In *Ecomposition: Theoretical and Pedagogical Approaches*. Edited by Christian R. Weisser and Sidney I. Dobrin. Albany: State University of New York Press, 2001.

———. "English 3310 Advanced Expository Writing: Rhetoric and Environment." *In Composiiton Studies* 27.2 (1999): 69–95.

———. "Paralogic Hermeneutic Theories, Power, and the Possibility for Liberating Pedagogies." In *Post-Process Theory: New Directions for Composition Research*. Edited by Thomas Kent, 145–83. Carbondale: Southern Illinois University Press, 1999.

———. *Constructing Knowledges: The Politics of Theory-Building and Pedagogy in Composition*. Albany: State University of New York Press, 1997.

———. "Race and the Public Intellectual: A Conversation with Michael Eric Dyson." *JAC: A Journal of Composition Theory* 17 (1997): 143–181.

Drew, Julie. "The Politics of Place: Student-Travelers and Pedagogical Maps." In *Ecomposition: Theoretical and Pedagogical Approaches*. Edited by Christian R. Weisser and Sidney I. Dobrin. Albany: State University of New York Press, 2001.

Duane, Daniel. *Caught Inside: A Surfer's Year on the California Coast*. New York: North Point Press, 1996.

During, Simon. *The Cultural Studies Reader*. New York: Routledge, 1999.

Eckersley, Robin. *Environmentalism and Political Theory: Towards an Ecocentric Approach*. Albany: State University of New York Press, 1992.

Egerton, F. N. "Ancient Sources for Animal Demography." *Isis* 59: 175–89.

Eisler, Riane. *The Chalice and the Blade*. San Francisco: Harper and Row, 1987.

Emerson, Ralph Waldo. *The Journals of Ralph Waldo Emerson*. Boston: Houghton Mifflin, 1909–1914.

Emig, Janet. "Writing as a Mode of Learning." In *College Composition and Communication* 28 (1977): 122–28. Reprinted in *Cross-Talk in Comp Theory: A Reader*. Edited by Victor Villanueva, Jr., 7–15. Urbana, IL: NCTE, 1997.

Evernden, Neil. "Beyond Ecology: Self, Place, and the Pathetic Fallacy." In *The Ecocriticism Reader: Landmarks in Literary Ecology*. By Cheryl Glotfelty and Harold Fromm, 92–104. Athens: University of Georgia Press, 1996.

———. *The Natural Alien*. Toronto: University of Toronto Press, 1985.

Faigley, Lester. *Fragments of Rationality: Postmodernity and the Subject of Composition*. Pittsburgh: University of Pittsburgh Press, 1992.

Finch, Robert, and John Elder, eds. *Nature Writing: The Tradition in English.* New York: W.W. Norton, 2002.

———. *The Norton Book of Nature Writing.* New York: W.W. Norton, 1990.

Fish, Stanley. *Professional Correctness: Literary Studies and Political Change.* New York: Oxford University Press, 1995.

Foucault, Michel. *Remarks on Marx: A Conversation with Duccio Trombadori.* R. James Goldstein and James Casciato, trans. New York: Sciotext (e), 1991.

Fox, Warwick. "The Deep Ecology—Ecofeminism Debate and Its Parallels." *Environmental Ethics* 11 (1989): 5–25.

Fraser, Nancy. "Rethinking the Public Sphere: A Contribution to the Critique of Actually Existing Democracy." In *Habermas and the Public Sphere.* Edited by Craig Calhoun. Cambridge, MA: MIT Press, 1996.

Gaard, Greta. "Living Interconnections with Animals and Nature." In *Ecofeminism: Women, Animals, Nature.* Edited by Greta Gaard, 1–12. Philadelphia: Temple University Press, 1993.

Gilligan, Carol. *In a Different Voice: Psychological Theory and Women's Development.* Cambridge: Harvard University Press, 1982.

Glotfelty, Cheryll, and Harold Fromm. *The Ecocriticism Reader: Landmarks in Literary Ecology.* Athens: University of Georgia Press, 1996.

Gore, Al. "Introduction," *Silent Spring.* Rachel Carson. Boston: Houghton Mifflin, 1994.

Gould, Steven Jay. *Eight Little Piggies: Reflections in Natural History.* New York: W. W. Norton, 1994.

Greenblatt, Stephen, and Giles Gunn, eds. *Redrawing the Boundaries: The Transformation of English and American Literary Studies.* New York: MLA, 1992.

Grossberg, Lawrence, Cary Nelson, Paula A. Treichler, eds. *Cultural Studies.* New York: Routledge and Kegan Paul, 1992.

Grey, Elizabeth Dodson, ed. *Sacred Dimensions of Women's Experience.* Wellesley, MA: Roundtable Press, 1988.

Guthrie, W. K. C. *The Sophists.* New York: Cambridge University Press, 1971.

Habermas, Jürgen. *The Structural Transformation of the Public Sphere.* Trans. T. Berger and F. Lawrence. Cambridge, MA: MIT Press, 1989.

Halden-Sullivan, Judith. "The Phenomenology of Process." In *Into the Field: Sites of Composition Studies.* Edited by Anne Ruggles Gere. New York: MLA, 1993.

Hall, Stuart. "Ethnicity: Identity and Difference." In *Becoming National: A Reader.* Edited by Geoff Eley and Ronald Grigor Suny. New York: Oxford University Press, 1996.

Halloran, Michael. "Rhetoric in the American College Curriculum: The Decline of Public Discourse." *Pre/text* 3 (1982): 245–69.

Halpern, David, and Dan Frank, eds. *The Nature Reader.* Hopewell, NJ: Ecco Press, 1997.

Haraway, Donna. *Simians, Cyborgs, and Women: The Reinvention of Nature.* New York: Routledge and Kegan Paul, 1991.

Harding, Sandra. *The Science Question in Feminism.* Ithaca: Cornell University Press, 1986.

———. *Whose Science? Whose Knowledge?: Thinking from Women's Lives.* Ithaca: Cornell University Press, 1991.

———. *Is Science Multicultural Postcolonialism, Feminism & Epistemologies: Postcolonialisms, Feminisms, and Epistemologies (Race, Gender, Science).* Bloomington: Indiana University Press, 1998.

Harkin, Patricia, and John Schilb, eds. *Contending with Words: Composition and Rhetoric in a Postmodern Age.* New York: MLA, 1991.

Harris, Joseph. *A Teaching Subject: Composition Since 1966.* Upper Saddle River, NJ: Prentice Hall, 1997.

Hawisher, Gail E., Paul LeBlanc, Charles Moran, Cynthia L. Selfe. *Computers and the Teaching of Writing in American Higher Education, 1979–1994: A History.* Norwood, NJ: Ablex, 1996.

Heilker, Paul. "Rhetoric Made Real: Civic Discourse and Writing Beyond the Curriculum." In *Writing the Community: Concepts and Models for Service-Learning in Composition.* Edited by Linda Adler-Kassner, Robert Crooks, and Ann Waters, 71–77. Urbana, IL: NCTE and AAHE, 1997.

Herndl, Carl G., and Stuart C. Brown, eds. *Green Culture: Environmental Rhetoric in Contemporary America.* Madison: University of Wisconsin Press, 1996.

Herzberg, Bruce. "Community Service and Critical Teaching." CCC 45 (1994): 307–19.

Hilbert, Betsy. "Teaching Nature Writing at a Community College." In *Teaching Environmental Literature: Materials, Methods, Resources.* By Frederick O. Waage. 88–92. New York: MLA, 1985.

Howarth, William. "Some Principles of Ecocriticism." In *The Ecocritism Reader: Landmarks in Literary Ecology.* By Cheryl Glotfelty and Harold Fromm. Athens: University of Georgia Press, 1996.

Hughes, J. Donald. "Early Greek and Roman Environmentalists." *Historical Ecology: Essays on Environment and Social Change.* Port Washington, NY: Kernnikat Press, 1980.

Ingram, Annie Merril. "Service learning and Ecocomposition: Developing Sustainable Practices through Inter- and Extradisciplinarity." In *Ecomposition: Theoretical and Pedagogical Approaches.* Edited by Christian R. Weisser and Sidney I. Dobrin. Albany: State University of New York Press, 2001.

Jagtenberg, Tom, and David McKie. *Eco-Impacts and the Greening of Postmodernity: New Maps for Communication Studies, Cultural Studies, and Sociology.* Thousand Oaks, CA: Sage Publications, 1997.

Jameson, Fredric. "Postmodernism, or the Cultural Logic of Late Capitalism." *New Leftist Review* 146 (1984): 71–72.

———. "Cognitive Mapping." In *Marxism and the Interpretation of Nature.* Cary Nelson and Lawrence Grossberg, eds. Urbana: University of Illinois Press (1988): 348.

Jarratt, Susan. *Rereading the Sophists: Classical Rhetoric Refigured.* Carbondale: Southern Illinois University Press, 1991.

Jenseth, Richard, and Edward E. Lotto, eds. *Constructing Nature: Readings from the American Experience.* Upper Saddle River, NJ: Prentice Hall, 1996.

Katz, Steven B., and Carolyn R. Miller. "The Low-Level Radioactive Waste Site Controversy in North Carolina: Toward a Rhetorical Model of Risk Communication." In *Green Culture: Environmental Rhetoric in Contemporary America.* Edited by Carl G. Herndl and Stuart C. Brown, 111–40. Madison: University of Wisconsin Press, 1996.

Keller, Christopher J. "Review of *Dramas of Solitude: Narratives of Retreat in American Nature Writing.*" *JAC* 19 (1999): 511–16.

Kennedy, George A. *Arisotle on Rhetoric: A Theory of Civil Discourse.* New York: Oxford University Press, 1991.

Kent, Thomas. "Externalism and the Production of Discourse." *JAC* 12 (1992): 57–94.

Kerferd G. B. *The Sophistic Movement.* New York: Cambridge University Press, 1981.

Killingsworth, M. Jimmie, and Jacqueline S. Palmer. *Ecospeak: Rhetoric and Environmental Politics in America.* Carbondale: Southern Illinois University Press, 1992.

Krebs, Charles J. *Ecology: The Experimental Analysis of Distribution and Abundance*. 2d ed. New York: Harper and Row, 1978.

Kuhn, Thomas S. *The Structure of Scientific Revolutions*. Chicago: University of Chicago Press, 1962.

Lindemann, Erika. "Three Views of English 101." *College English:* 57.3 (1995).

Lopez, Barry. "Barry Lopez." *On Nature*. Edited by Daniel Halpern, 295–97. San Francisco: North Point Press, 1987.

———. "Searching for Depth in Bonaire." In *American Nature Writing*, 1998. Edited by John A. Murray, 12–28. San Francisco: Sierra Club Books, 1998.

Lord, Nancy. *Fish Camp: Life on an Alaskan Shore*. Washington, DC: Counterpoint, 1997.

Lyons, Nick. *My Secret Fishing Life*. New York: Atlantic Monthly Press, 1999.

Macauley, David. "Greening Philosophy and Democratizing Ecology." In *Minding Nature: The Philosophers of Ecology*. Edited by David Macauley, 1–23. New York: Guilford Publications, 1996.

Macauley, David, ed. *Minding Nature: The Philosophers of Ecology*. New York: Guilford Publications, 1996.

Manes, Christopher. "Nature and Silence." *The Ecocriticsm Reader: Landmarks in Literary Ecology*. Edited by Cheryl Glotfelty and Harold Fromm, 15–29. Athens: University of Georgia Press, 1996.

Marshall, P. *Nature's Web: An Exploration of Ecological Thinking*. London: Simon and Schuster, 1992.

McAndrews, Donald A. "Ecofeminism and the Teaching of Literacy." CCC 47 (1996): 367–382.

McDowell, Michael. "Edge Effects: Exploring Natural and Discursive 'Places' through Computer-Assisted Ecocomposition." Paper presented at Convention of the Conference on College Composition and Communication. Chicago, April, 1998.

———. "Talking about Trees in Stumptown: Pedagogical Problems in Teaching EcoComp." *Reading the Earth: New Directions in the Study of Literature and the Environment*. Edited by Michael P. Branch, Rochelle Johnson, Daniel Patterson, and Scott Slovic, 19–28. Moscow: University of Idaho Press, 1998.

McFadden, Margaret. "'The I in Nature': Nature Writing as Self-Discovery." In *Teaching Environmental Literature: Materials, Methods, Resources*. By Frederrick O. Waage, 102–107. New York: MLA, 1985.

Merleau-Ponty, Maurice. *The Visible and the Invisible*. Evanston, IL: Northwestern University Press, 1968.

Merod, Jim. *The Political Responsibility of the Critic*. Ithaca: Cornell University Press, 1987.

Miller, Susan. "The Feminization of Composition." In *Composition in Four Keys*. Edited by Mark Wiley, Barbara Gleason, Louise Wetherbee Phelps, 492–502. Mountain View, CA: Mayfield Publishing Company, 1996.

———."Technologies of Self?-Formation." *JAC: A Journal of Composition Theory* 17.3 (1997): 497–500.

Monsma, Bradley. "Writing Home: Composition, Campus Ecology, and Webbed Environments. In *Ecocompositiom: Theortical and Pedagogical Approaches*. Edited by Christian R. Weisser and Stanley I. Dobrin. Albany: State University of New York, 2001.

Mumford, Lewis. *The Story of Utopias*. New York: Viking Press, 1950.

Myers, Greg. "The Social Construction of Two Biological Proposals." *Written Communication* 2 (1985): 219–45.

Nash, Roderick. *The Rights of Nature: A History of Environmental Ethics*. Madison: University of Wisconsin Press, 1989.

Negt, Oscar, and Alexander Kluge. *The Public Sphere and Experience*. Minneapolis, MN: University of Minnesota Press, 1996.

Nelson, Cary, Paula A. Treichler, and Lawrence Grossberg. "Cultural Studies: An Introduction." In *Cultural Studies*. Edited by Lawrence Grossberg, Cary Nelson, and Paula A. Treichler, 1–22. New York: Routledge and Kegan Paul, 1992.

"New CCCC Service-Learning Committee Formed." *The Council Chronicle* 9 (Nov, 1999): 6.

Ohman, Richard. *Politics of Letters*. Middletown, CT: Wesleyan University Press, 1987.

Olson, Gary A. "History, *Praxis*, and Change: Paulo Freire and the Politics of Literacy." *JAC* 12 (1992): 1–14.

Olson, Gary A., and Lester Faigley. "Language, Politics, and Composition: A Conversation with Noam Chomsky." *JAC* 11 (1991): 1–35.

Owens, Derek. *Composition and Sustainability: Teaching for a Threatened Generation*. Urbana, IL: NCTE, 2001.

Pirsig, Robert. *Zen and the Art of Motorcycle Maintenance: An Inquiry into Values*. William Morrow, 1974.

Plevin, Arlene. "The Liberatory Positioning of Place in Ecocomposition: Reconsidering Paulo Freire." In *Ecocmposition: Theoretical and Pedagogical Approaches*. Edited by Christian R. Weisser and Sidney I. Dobrin. Albany: State University of New York Press, 2001.

Plumwood, Val. "Nature, Self, and Gender: Feminism, Environmental Philosophy, and the Critique of Rationalism." *Hypatia* 6 (1991): 3–27.

Porter, James E. *Audience and Rhetoric: An Archaeological Composition of the Discourse Community*. Upper Saddle River, NJ: Prentice Hall, 1997.

Ricklefs, Robert E. *The Economy of Nature*. 4th ed. New York: W. H. Freeman and Company, 1997.

Rifkin, James. *Biosphere Politics: A New Consciousness for a New Century*. New York: Crown Press, 1991.

Robertson, Linda R., Sharon Crowley, and Frank Lentricchia. "The Wyoming Conference Resolution Opposing Unfair Salaries and Working Conditions for Post-Secondary Teachers of Writing." *College English* 49 (1987): 274–80.

Roorda, Randall. "Process and Product in Place: Qualitative Natural History and Ecological Literacy in Composition." Paper presented at the Convention of the Conference on College Composition and Communication. Chicago, April, 1998.

——. "Nature/Writing: Literature, Ecology, and Composition." *JAC: A Journal of Composition Theory* 17 (1997): 401–414.

——. *Dramas of Solitude: Narratives of Retreat in American Nature Writing*. Albany: State University of New York Press, 1998.

Rose, Mike. "The Language of Exclusion: Writing Instruction at the University." *College English* 47.4 (1985): 341–59.

Russell, David R. "Vygotsky, Dewey, and Externalism: Beyond the Student/Disciple Dichotomy." *JAC* 13 (1993): 504–54.

Sanchez, Raul. Postcolonial@jefferson.village.virginia.edu. 14 September 1997.

Sanders, Scott Russell. *Writing from the Center*. Bloomington: Indiana University Press, 1995.

Sarles, Harvey. *Language and Human Nature*. Minneapolis, MN: University of Minnesota Press, 1985.

Scholes, Robert. *Textual Power: Literacy Theory and the Teaching of English*. New Haven: Yale University Press, 1985.

Slicer, Deborah. "Your Daughter or Your Dog? A Feminist Assessment of the Animal Research Issue." *Hypatia* 6 (1991): 108–24.

Slovic, Scott. "Ecocriticism: Storytelling, Values, Communication, Contact." http://wsrv. clas. virginia.edu/~djp2n/conf/WLA/slovic.html. 8 Dec. 1998, 2:30 pm.

———. "Nature Writing and Environmental Psychology: The Interiority of Outdoor Experience." In *The Ecocriticism Reader: Landmarks in Literary Ecology. The Ecocriticism Redaer: Landmarks in Literary Ecology*. By Cheryl Glotfelty and Harold Fromm. 351–70. Athens: University of Georgia Press, 1996.

Smith, Lee. "Teaching Ecocomposition in the Urban University." Convention of the Conference on College Composition and Communication. Chicago, April, 1998.

Gary Snyder. *The Practice of the Wild*. San Francisco: North Point Press, 1990.

Snow, C. P. *The Two Cultures and the Scientific Revolution*. New York: Cambridge University Press, 1959.

Strauss, Leo. *Natural Right and History*. Chicago: University of Chicago Press, 1953.

Sumner, David. "Don't Forget to Argue: Problems, Possibilities, and Ecocomposition." In *Ecomposition: Theoretical and Pedagogical Approaches*. Edited by Christian R. Weisser and Sideny I. Dobrin. Albany: State University of New York Press, 2001.

Swan, James A. *In Defense of Hunting*. New York: Harper Collins, 1995.

Tag, Stan. "Four Ways of Looking at Ecocriticism." Http://wsrv.clas.virginia. edu/~djp2n/ conf/WLA/tag.html. 8 Dec. 1998, 2:30 pm.

Turner, Graeme. *British Cultural Studies: An Introduction*. New York: Routledge and Kegan Paul, 1990.

Ulman, H. Lewis. "Thinking Like a Mountain: Persona, Ethos, and Judgment in American Nature Writing." In *Green Culture: Environmental Rhetoric in Contemporary America*. Carl G. Herndl and Stuart C. Brown, eds. Madison: University of Wisconsin Press, 1996.

Verburg, Carol J. *The Environmental Predicament: Four Issues for Critical Analysis*. New York: Bedford/St. Martin's, 1995.

Villanueva, Victor, Jr. "Considerations for American Freireistas." *The Politics of Writing Instruction: Postsecondary*. Edited by Richard Bullock and John Trimbur. Portsmouth, NH: Boynton/Cook, 1991.

Waage, Frederick O. *Teaching Environmental Literature: Materials, Methods, Resources*. New York: MLA, 1985.

Walker, Alice. *Living by the Word*. San Diego: Harcourt Brace Jovanovich. 1989.

Ward, Irene. "How Democratic Can We Get?: The Internet, the Public Sphere, and Public Discourse." *Literacy, Ideology, and Dialogue: Toward a Dialogic Pedagogy*. Albany: State University of New York Press, 1994.

Warren, Karen J. "Feminism and Ecology: Making Connections." *Environmental Ethics* 9 (1987): 3–20.

———. "Ecological Feminist Philosophies: An Overview of the Issues." *Ecological Feminist Philosophies*. Edited by Karen J Warren, ix–xxvi. Bloomington: Indiana University Press, 1996.

———. *Ecological Feminist Philosophies*. Bloomington: Indiana University Press, 1996.

———. *Ecofeminism: Women, Culture, Nature.* Bloomington: Indiana University Press, 1997.

Watkins, Evan. *Worktime: English Departments and the Circulation of Cultural Value.* Stanford: Stanford University Press, 1989.

Weisser, Christian R. "Campus Ecology: Environmental Issues on the University of Tampa's Campus." http://utweb.utampa.edu/faculty/cweisser/campus.htm. 8 Dec 1999, 2:30 pm.

———. "Discourse and Authority in Electronic Contact Zones." *The Writing Instructor* 16.3 (1997): 103–12.

———. "Ecocomposition and the Greening of Identity." In *Ecocomposition: Theoretical and Pedagogical Approaches.* Edited by Christian R. Weisser and Sidney I. Dobrin. Albany: State University of New York Press, 2001.

Weisser, Christian R., and Sidney I. Dobrin, eds. *Ecocomposition: Theoretical and Pedagogical Approaches.* Albany: State University of New York Press, 2001.

Wells, Susan. "Rogue Cops and Health Care: What Do We Want from Public Writing? CCC 47 (1996): 325–41.

White, Daniel R. *Postmodern Ecology: Communication, Evolution, and Play.* Albany: State University of New York Press, 1998.

White, Edward M. *Assigning, Responding, Evaluating: A Writing Teacher's Guide.* 3d ed. New York: St. Martin's Press, 1995.

Williams, Terry Tempest. "The Erotic Landscape." *American Nature Writing 1995.* Edited by John A. Murray. San Francisco: Sierra Club Books, 1995.

Wilson, Edward O. *Consilience: The Unity of Knowledge.* New York: Vintage Books, 1998.

Worsham, Lynn. "Writing against Writing: The Predicament of *Ecriture Feminine* in Composition." In *Contending with Words: Composition and Rhetoric in a Postmodern Age.* Edited by Patricia Harkin and John Schilb, 82–104. New York: Modern Language Association, 1991.

Worster, Donald. *Nature's Economy: A History of Ecological Ideas.* Cambridge: Cambridge University Press, 1977.

Wright, Derek. "Parenting the Nation: Some Observations on Nuruddin Farah's *Maps.*" In *Order and Partialities: Theory, Pedagogy, and the 'Postcolonial.'* Kostas Mysiades and Jerry McGuire, eds. Albany: State University of New York Press, 1985.

Yaeger, Patricia. *The Geography of Identity.* Ann Arbor: University of Michigan Press, 1996.

Index